Reptiles and Amphibians
of the Pacific Islands

Reptiles and Amphibians of the Pacific Islands

A Comprehensive Guide

George R. Zug

UNIVERSITY OF CALIFORNIA PRESS

Berkeley · Los Angeles · London

University of California Press, one of the most distinguished university presses in the United States,
enriches lives around the world by advancing scholar-
ship in the humanities, social sciences, and natural
sciences. Its activities are supported by the UC Press
Foundation and by philanthropic contributions from
individuals and institutions. For more information,
visit www.ucpress.edu.

University of California Press
Berkeley and Los Angeles, California

University of California Press, Ltd.
London, England

Library of Congress Cataloging-in-Publication Data

Zug, George R., 1938-
   Reptiles and amphibians of the Pacific Islands :
a comprehensive guide / George R. Zug.
      pages   cm
   Includes bibliographical references and index.
   ISBN 978-0-520-27495-2 (cloth : alk. paper) —
   ISBN 978-0-520-27496-9 (pbk. : alk. paper)
   1. Reptiles—Islands of the Pacific—Identification.
2. Amphibians—Islands of the Pacific—Identification.
I. Title.
   QL664.A1Z84   2013
   597.909—dc23   2012037421

Manufactured in China

22   21   20   19   18   17   16   15   14   13
10   9   8   7   6   5   4   3   2   1

The paper used in this publication meets the minimum
requirements of ANSI/NISO Z39.48-1992 (R 2002)
(*Permanence of Paper*). ⊗

In gratitude to the many Pacific Islanders who aided my research and to past and present colleagues who have contributed to our knowledge of Pacific island herpetofaunas.

Contents

IDENTIFICATION PLATES: BETWEEN PAGES 54 AND 55.

Preface

Amid the vastness of the Pacific Ocean, tiny specks of land have served as evolutionary cauldrons, generating enormous variety in the species that inhabit them. This species diversity is generally unrecognized by Pacific island residents and visitors. Only since the late 1970s have systematic biologists begun to focus on this diversity, breaking with the earlier assumption that the faunas and floras of the widely distant islands consisted predominantly of shared species. This assumption of shared faunas was especially prominent among herpetologists until Pacific biologists such as Dick Watling and John Gibbons began to look more closely at their local herpetofaunas. They and their colleagues discovered substantial local and regional differentiation. The use of new molecular tools, in concert with traditional morphological techniques, allows us to recognize that populations of similar morphologies are genetically distinct; often with the genetic data, we discover that the morphology of different populations is not as similar as originally supposed. The pace of new species recognition has quickened with the use of these molecular tools, and, excitingly, we have discovered that relationships among congeners show complex dispersal histories.

My participation in some of these discoveries, coinciding with the arrival of an increasing number of alien species, along with the absence of a Pacific-wide guide to reptiles and amphibians, persuades me to offer this summary of the tropical Pacific islands' herpetofaunas in the form of an identification guide. Although I hope that professional herpetologists will find new data and the summaries useful, they are not my main audience. My audience is the casual to avid naturalists who live on or visit Pacific islands and are curious about the lizards and other herps that they see daily.

Pacific islands, from my perspective, are largely those islands that are considered part of Oceania. However, one cannot refer to "Pacific islands" and ignore the Galápagos Islands and other eastern Pacific islands. For the purposes of this guide, I am concerned with the herpetofaunas of only oceanic islands—that is, islands with no past land (subaerial) connections to a continental land mass. Oceanic islands are those islands that have always been surrounded by seawater. Amphibians and reptiles colonize these islands by cross-water dispersal, although not necessarily in or on the water. The boundaries for the Pacific islands herein are latitudinally 30° north to 30° south and longitudinally 130° east to 90° west. These boundaries encompass the Hawaiian and Bonin Islands in the north and Easter Island in the south (Norfolk Island and Kermadec Islands excluded), and the Palau Islands in the west to the Galápagos Islands in the east. I exclude the herpetofauna of New Caledonia because this land mass is an ancient fragment of Australia-Gondwana and its fauna has only a smattering of typical Pacific species.

Identification of an organism commonly relies on geography; thus consulting a series of regional herpetofaunal lists is the first step in identification. These regional checklists identify some centers of diversity, but they provide no explanation for the speciation within and among the oceanic herpetofaunas of the Pacific islands. With knowledge of the possible species, the user can compare the animal in hand with the images in the identification plates and confirm the tentative pictorial identification with the morphological details presented in the species accounts. These accounts also offer a modicum of information on the biology of each species.

Introduction

The Pacific Ocean is a remnant of the once great sea that surrounded the supercontinent Pangaea. As Pangaea broke up and its fragments began to move apart, the Indian and the Atlantic Oceans developed. Although the Pacific Ocean is only a shadow of its former self, it remains the largest of our present oceans and covers about a third of the Earth's surface, roughly 180 million square kilometers (about 70 million square miles).

The floor of the Pacific Ocean lies at an average depth of 4,000 meters, with its deepest spot—more than 11 kilometers (36,198 ft)—found in the Mariana Trench in the western Pacific. The ocean floor is neither smooth nor a gently sloping basin. Thousands of seamounts are scattered throughout; some of these extend above the sea's surface as islands or groups of islands. The ocean floor also contains plateaus, ridges, and trenches. All these features derive from tectonic activity within and at edges of the oceanic plates.

The Earth's crust is comprised of oceanic and continental plates abutting one another, with no gaps between. Each plate consists of the lithosphere and upper ductile portion of the Earth's mantle and floats on the deeper, more solid portion of the mantle. Continental plates are thicker (30–70 km) but lighter and float higher than the more dense oceanic plates (which are 3–7 km thick). The plates move because one or more of their edges is a growth zone. Growth results from a constant upwelling of molten material from the underlying mantle. As the upwelling adds new material to the edges of the abutting plates, the older portions of the plates move away from the growth zone. Because the molten material is hotter than the older portion of the plates, it is less dense and rises above the ocean floor, creating oceanic ridges or rises. These oceanic ridges are divergent zones, because the adjacent plates are slowly growing and moving away from one another. The opposite edge of each plate, where it abuts against another plate, is called the "convergent zone." The heavier oceanic plate pushes beneath the thicker, but lighter, continental plate or, if two oceanic plates collide, both move downward. At these convergent zones the oceanic plate margins subduct (move downward) into the mantle.

On one side of a subduction zone, volcanic activity and plate uplift usually create either mountain chains (if the zone is on the edge of a continental plate) or island archipelagos (if the zone is on an oceanic plate). Islands also arise within oceanic plates where a plume of mantle pushes through the plate and erupts as a volcano on the plate's surface. These submarine volcanoes commonly persist for centuries. Their periodic lava outflows gradually build a mound that eventually rises above the sea's surface to form an island. These mantle hotspots are relatively stationary within the lower mantle; as the plate continues to move, the volcano is eventually cut off from the mantle plume and loses its magma source. Soon, geologically speaking, another volcano arises and begins to build toward the surface of the sea. The weathering of the extinct volcano and the cooling and subsiding of the adjacent plate slowly erode and drown the island, which becomes a seamount.

The plate's movement over a hotspot creates a linear chain or arc of islands. The oldest islands in the chain are the most distant from the hotspot and also the smallest, because they have undergone the longest period of erosion. The Hawaiian Islands, for example, comprise an archipelago of hotspot islands on the Pacific Plate; the Island of Hawai'i is still over the hotspot and continuing to grow; in the distant northwest of the archipelago, the French Frigate Shoals are sinking below sea level and becoming seamounts. To take another example, the Galápagos Islands are formed from a hotspot on the Nazca Plate. Other islands and island chains form along the convergent zones. Intense interactions between subducting plates and the mantle often result in strings of volcanoes beside the deep abyssal trenches; the Mariana Archipelago was created in this manner. Subduction zones and rises also shift.

These geological changes have occurred regularly within the Southwest Pacific; the Tongan and Fijian islands were created along such convergent zones that are no longer active.

Pacific islands have repeatedly formed from the processes outlined above, although in a much more complex manner than described here. The islands and island chains that we see today have been around only relatively briefly—generally no more than 40 million years. During that short interval, islands have grown and disappeared; island chains have rotated, changing their positions relative to nearby chains. The arrival and colonization of these islands by reptiles began soon after plants arrived and established a protective and nourishing cover for animal life. Exactly when the present lineages arrived on each island group is only beginning to be discovered. The complex timing and sequence of arrival varies greatly across the great expanse of the Pacific. We can expect a variety of evolutionary scenarios. One scenario is seen in the Galápagos, where the oldest island dates to about 4–5 million years ago, yet data from molecular genetics indicate that some lineages of Galápagos organisms began diverging from their mainland ancestors 8–10 million years ago and apparently arrived on a hotspot island that is now only a seamount.

THE CONTEMPORARY PACIFIC

The Pacific is bound on the north by the Bering Straits; on the east by the Americas; on the west by the continents of Asia, Australia, and the islands on the platform between them; and on the south it merges without a clear surface demarcation into the Antarctic Ocean. This enormous area encompasses numerous nations and island groups containing an estimate of over 24,000 separate islands.

What is and is not part of the Pacific—particularly the western Pacific—is variously delimited. Some authorities have the Pacific and the Indian Oceans abutting the western edge of the Lesser and Greater Sunda Islands. Other authorities set the western edge at the eastern edge of this Sundan platform. A majority viewpoint accepts the landmasses of Japan, Taiwan, the Philippines, New Guinea, and eastern Australia, which face the open waters of the Pacific, as its western edge.

A broad array of regions, islands, island groups, and nations are found within the Pacific. The simplest grouping to describe is the distinction between continental and oceanic islands. Continental islands have had a subaerial (above water or sea level) connection to a continental landmass. They lie on the shallower portions of the continental shelf and are largely found in the western Pacific because the shelf in the eastern Pacific is nearly nonexistent and the continental slope of the Americas is extremely steep. Only a few islands lie on this narrow eastern Pacific shelf; they are very close to the American mainland and include the Mexican Tres Marías, the California Channel Islands, the Queen Charlotte Islands, and the Aleutian Islands. In contrast, the western Pacific Rim islands of Japan, the Ryukyu Islands, Taiwan, the Philippines, New Guinea, and the Great Barrier Reef islands are all continental islands.

Some additional regional terms apply largely to oceanic islands. The most obvious one is "Oceania." A standard definition of Oceania includes Australia, Papua New Guinea, the Solomon Islands, New Caledonia, and New Zealand and the oceanic islands of Polynesia, Micronesia, and Melanesia. This definition of Oceania excludes the Philippines, Japan, eastern Pacific islands (e.g., the Galápagos Islands, Clipperton Island), and even the Hawaiian Islands. My preferred usage stresses the oceanic part of Oceania and includes all the islands of Micronesia and Polynesia as well as the oceanic part of Melanesia westward into the eastern Solomon Islands. The Australopapuan landmass is continental. New Caledonia and New Zealand are excluded because they are fragments of the supercontinent Pangaea.

HUMAN HISTORY IN THE PACIFIC

Melanesia, Micronesia, and Polynesia are ethnographic regions defined by the cultures and origins of their human inhabitants. These are well-established, useful geographic terms. "Melanesia" is comprised of New Guinea and the groups of islands west of it to Fiji. "Micronesia" lies north of the equator and includes the widely scattered islands from the Palau-Marianas axis to the International Dateline. "Polynesia" covers a vast area of the Central Pacific east of the International Dateline from the Hawaiian Islands southward to Rapa Nui (Easter Island). A westward tongue of Polynesia extends diagonally across the Dateline from Tonga to Nauru.

Melanesia was the first area settled by humans among the Pacific islands, with people arriving in New Guinea at least by 40,000 BCE. These people, who became the Papuans, were largely hunters and gatherers, dependent on the plants and animals of the forest and nearshore marine environments. They were able to cross narrow water gaps with crude canoes and rafts. Presumably they attempted a crossing only when they were able to see the land mass of their destination. The multiplicity of Papuan languages attests to multiple invasions of New Guinea by peoples of many different cultures and at different times.

A diaspora of people from Taiwan and adjacent mainland China began about 3000 BCE. These people, now known as "Austronesians," were the colonizers of Oceania. Austronesians were horticulturalists. They carried domesticated plants (e.g., yams, taro) and animals (e.g., pigs, dogs, chickens) in their settlement of new lands. Their double-outrigger canoes allowed them to transport more people and cargo, and permitted travel to more distant lands that were beyond their sight-horizon. The latter ability suggests the possession of elementary navigation skills and the identification of suitable settlement sites by previous exploration.

The Austronesians spread southward into the Philippines, which they settled, displacing the earlier arriving Negritos populations. Their dispersal continued, with groups moving westward into the Sunda Islands, southward to northern New Guinea and the Bismarck Archipelago, and from there eastward into southern Oceania and eastward into the Palau and Mariana Archipelagos, eventually populating most of western Micronesia.

The Austronesians moving along New Guinea were coastal inhabitants, mostly living in stilted villages over water. They certainly interacted with the Papuan populations, introducing the latter to their domesticated crops and animals, tools, and ceramics. This Austronesian culture became known as the "Lapita." The Lapita continued to move eastward, reaching easternmost Solomon Islands by about 1300 BCE, Fiji by about 1100 BCE, and the Tongan and Samoan islands by about 900 BCE. Migration paused, although interchange among the peoples of these two latter island groups continued and slowly the new Polynesian culture evolved. In the middle of the first millennium AD, apparently a new major phase of dispersal resulted in the population of the remainder of Oceania. This time, large numbers of individuals (40 to 60) and their cargo were able to travel to other islands in large double-hulled canoes. These canoes were a major development of the proto-Polynesian culture. The Cook Islands, the Marquesas and Society Islands, and the Hawaiian Islands were settled by this method by 800 AD. New Zealand was the last major island reached by the Polynesians, who settled there in the 1200s AD.

Plants and animals had been scattered throughout the Pacific long before humans arrived. Plants, quickly followed by animals, arrived soon after land rose above the sea. Biological communities that developed over many millennia were drastically disrupted, their component species decimated, and many driven extinct with arrival of the Austronesians. Present day island floras and faunas are vastly different from the prehuman ones. In many instances, the floras and faunas are now dominated by alien species whose prehistoric presence has made them appear as though they are natives. This guide addresses only the reptiles and amphibians inhabiting the oceanic Pacific islands today; the origin and dispersal of the Pacific island herpetofaunas are not addressed.

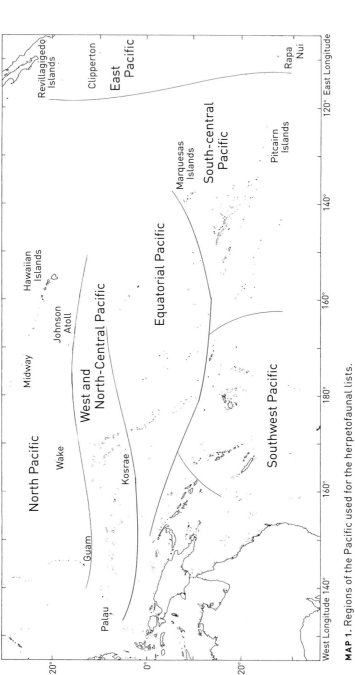

MAP 1. Regions of the Pacific used for the herpetofaunal lists.

Island and Island Group Herpetofaunas

My division and arrangement of the islands and island groups of the tropical Pacific are largely arbitrary. The herpetofaunal lists are presented north to south and sweep in broad latitudinal bands from west to east. These bands are arranged to reflect the principal direction in which the Pacific islands were colonized, with the exception of the East Pacific islands near the coastline of the Americas. The *North Pacific* band includes six island groups: the Bonin and Volcano Islands (Ogasawara Archipelago); the Hawaiian Islands; the Northern Mariana Islands (includes Rota); Guam; Wake Island; and Johnston Atoll. The *West- and North-Central* band encompasses five areas: Palau; Micronesia-Yap; Micronesia-Chuuk; Micronesia-Pohnpei (includes Kosrae); and the Marshall Islands (the Ralik and Ratok chains). The *Equatorial* band includes nine island groups: Kapingamarangi Atoll; Nauru; western Kiribati (the Banaba and "Tarawa" chains); the Howland and Baker islands; the Kiribati-northern Line Islands (includes Palmyra Atoll, Jarvis Island, and Kirimati); Tuvalu; Kiribati-Rawaki; Tokelau; and the Kiribati-southern Line Islands. The *Southwest Pacific* contains eight island clusters: Santa Cruz and the eastern outer Solomon Islands; Vanuatu; Rotuma; Fiji; Tonga; Wallis and Futuna; Samoa (includes American Samoa); and Niue. The *South-Central* band comprises Polynesia and nearby island groups: the Cook Islands; the Austral and Society Islands; the Tuamotu Archipelago and the Gambier Islands; the Marquesas Islands; and the Pitcairn Islands. The final group, the *Eastern Pacific*, contains the islands near the Americas: Islas Revillagigedo; Clipperton Island; Isla del Coco; Isla Malpelo; the Galápagos Archipelago; Rapa Nui or Easter Island; and Islas Sala y Gómez.

I have not identified faunas for individual islands within the larger groupings other than for endemic species that occur on one or a few islands; in these instances, the specific islands are identified within square brackets following the endemic species. All species identified for an island group do not necessarily, or even probably, occur on each island within the group. A species in an island group listing is not included if I have not located a museum or literature voucher for it. Such absences from an island list are most evident among the seaturtles.

Scientific names for amphibians and reptiles generally follow the nomenclature provided at two websites. The database for amphibians is the *Amphibian Species of the World: An Online Reference,* available at http://research.amnh.org/vz/herpetology/amphibia/. The database for reptiles is the *Reptile Database,* available at http://www .reptile-database.org. The former is updated regularly, the latter less regularly, hence my nomenclature choices might deviate from those in the databases. Also the pace of systematic studies and publication has quickened recently, so it is likely that new species will be described and nomenclatural analyses will alter some names between the completion of the manuscript and its publication.

Since the beginning of the 21st century, herpetologists and conservationists have made an effort to standardize the English common names of the world's amphibians and reptiles. Colloquial or common names vary with local language and tradition. The use of standard English names represents a relatively recent trend, seemingly begun with Conant's and Stebbin's field guides and then formalized by the North American herpetological societies' periodically revised list of names.* I follow their most recent recommendation for North American taxa and names that have become standardized by frequent usage in the herpetocultural and herpetological literature.

---

*The European herpetological community, especially the French and Germans, preceded the North American one with the development of standardized names for amphibians and reptiles.

For standard names, I decided to use island names as nouns rather than attempt to make them into adjectives because made-up adjectives—such as "Yapan" and "Guaman"—can be strange and ill-sounding. I retain widely used standard names, even when I dislike them—for example, Mourning Gecko. However, I prefer and recommend standard names that highlight a visible characteristic or geographic occurrence of a taxon, and

I use the term "alien" to denote taxa that have arrived during the last century or so and have established breeding populations. Alien arrivals are human-assisted, whether intentionally or accidentally. Invasive species are alien ones that have become well established, abundant, and disruptive of the previously existing biological communities. Many species—such as the Pacific Moth Skink—also owe their widespread distribution in Oceania to human-assisted transport, but the precise time when the dispersal of this species occurred has been difficult to determine. The timing of dispersal for species and populations that arose from natural dispersal is equally difficult to determine.

NORTH PACIFIC

Bonin and Volcano Islands [Ogasawara Guntö and Kazan Rettö]

FROGS (ANURA)

Rhinella marina	Cane Toad
Lithobates catesbeianus	American Bullfrog

LIZARDS (SQUAMATA)

Hemidactylus frenatus	Common House Gecko
Lepidodactylus lugubris	Mourning Gecko
Perochirus ateles	Micronesia Saw-tailed Gecko
Anolis carolinensis	Green Anole
Cryptoblepharus nigropunctatus	Bonin Snake-eyed Skink

SNAKES (SQUAMATA: SERPENTES)

Ramphotyphlops braminus	Brahminy Blindsnake

TURTLES (TESTUDINES)

Caretta caretta	Loggerhead Seaturtle
Chelonia mydas	Green Seaturtle

hence I replace the surnames of discoverers in names of Pacific taxa with place names or descriptors—for example, I use "Micronesia Spotted Skink" instead of "Boettger's Skink." In the tradition of combined English names, such as rattlesnake and blindsnake, I regularly combine double names—for example, seaturtle, treeskink, seasnake—when they denote a morphological, habitat, or behavioral group. Some insular populations are recognized as distinct species, although they have not yet been formally described; these undescribed species are identified by the symbol nsp (= new species).

Eretmochelys imbricata	Hawksbill Seaturtle
Dermochelys coriacea	Leatherback Seaturtle
Trachemys scripta	Red-eared Slider
CROCODILES (CROCODILIA)	NONE

Hawaiian Islands [Midway to Hawai'i]

FROGS (ANURA)

Rhinella marina	Cane Toad
Dendrobates auratus	Green and Black Poison-dart Frog
Eleutherodactylus coqui	Coquí
Eleutherodactylus planirostris	Greenhouse Frog
Glandirana rugosa	Japanese Wrinkled Frog
Lithobates catesbeianus	American Bullfrog

LIZARDS (SQUAMATA)

Chamaeleo calyptratus	Veiled Chameleon
Trioceros jacksonii	Jackson's Chameleon
Gehyra insulensis	Pacific Stump-toed Gecko
Gekko gecko	Tokay
Hemidactylus frenatus	Common House Gecko
Hemidactylus garnotii	Fox Gecko
Hemiphyllodactylus typus	Indo-Pacific Slender Gecko
Lepidodactylus lugubris	Mourning Gecko
Phelsuma grandis	Madagascar Giant Daygecko
Phelsuma guimbeaui	Orange-spotted Daygecko
Phelsuma laticauda	Golddust Daygecko
Anolis carolinensis	Green Anole
Anolis equestris	Knight Anole
Anolis sagrei	Brown Anole
Iguana iguana	Green Iguana
Cryptoblepharus poecilopleurus	Oceania Snake-eyed Skink

Emoia cyanura	White-bellied Copper-striped Skink [Extinct]
Emoia impar	Dark-bellied Copper-striped Skink
Lampropholis delicata	Garden Skink
Lipinia noctua	Pacific Moth Skink

SNAKES (SQUAMATA: SERPENTES)

Ramphotyphlops braminus	Brahminy Blindsnake
Pelamis platura	Yellow-bellied Seasnake

TURTLES (TESTUDINES)

Caretta caretta	Loggerhead Seaturtle
Chelonia mydas	Green Seaturtle
Eretmochelys imbricata	Hawksbill Seaturtle
Lepidochelys olivacea	Olive Ridley Seaturtle
Dermochelys coriacea	Leatherback Seaturtle
Trachemys scripta	Red-eared Slider
Apalone spinifera	Spiny Softshell
Palea steindachneri	Wattle-necked Softshell
Pelodiscus sinensis	Chinese Softshell

CROCODILES (CROCODILIA)   NONE

Northern Mariana Islands [Rota and northward]

FROGS (ANURA)

Rhinella marina	Cane Toad

LIZARDS (SQUAMATA)

Gehyra insulensis	Pacific Stump-toed Gecko
Gehyra oceanica	Oceania Gecko
Hemidactylus frenatus	Common House Gecko
Lepidodactylus lugubris	Mourning Gecko
Nactus pelagicus	Pacific Slender-toed Gecko
Perochirus ateles	Micronesia Saw-tailed Gecko

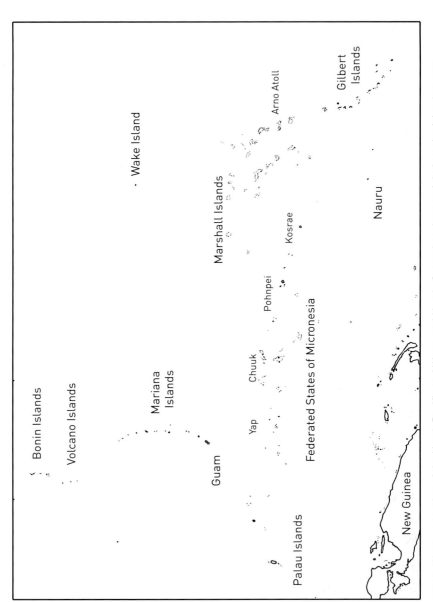

MAP 2. Western portion of the North Pacific and West- and North-Central regions and their major island groups.

Anolis carolinensis	Green Anole
Carlia ailanpalai	Admiralty Brown Skink
Cryptoblepharus poecilopleurus	Oceania Snake-eyed Skink
Emoia atrocostata	Seaside Skink
Emoia caeruleocauda	Pacific Blue-tailed Skink
Emoia slevini	Mariana Skink
Lamprolepis smaragdina	Emerald Treeskink
Varanus "indicus"	Pacific Monitor

SNAKES (SQUAMATA: SERPENTES)

Ramphotyphlops braminus	Brahminy Blindsnake

TURTLES (TESTUDINES)

Chelonia mydas	Green Seaturtle
Eretmochelys imbricata	Hawksbill Seaturtle
Trachemys scripta	Red-eared Slider

CROCODILES (CROCODILIA)   NONE

Guam

FROGS (ANURA)

Rhinella marina	Cane Toad
Fejervarya cancrivora	Crab-eating Frog
Fejervarya "limnocharis"	Paddy Frog
Eleutherodactylus planirostris	Greenhouse Frog
Hylarana guentheri	Brown and Tan Amoy Frog
Microhyla pulchra	Marbled Pygmy Frog
Litoria fallax	Eastern Dwarf Treefrog
Polypedates braueri	Taiwan Whipping Frog

LIZARDS (SQUAMATA)

Gehyra insulensis	Pacific Stump-toed Gecko
Gehyra oceanica	Oceania Gecko
Hemidactylus frenatus	Common House Gecko
Lepidodactylus lugubris	Mourning Gecko

Nactus pelagicus	Pacific Slender-toed Gecko
Anolis carolinensis	Green Anole
Perochirus ateles	Micronesia Saw-tailed Gecko [Extinct]
Carlia ailanpalai	Admiralty Brown Skink
Cryptoblepharus poecilopleurus	Oceania Snake-eyed Skink
Emoia atrocostata	Seaside Skink
Emoia caeruleocauda	Pacific Blue-tailed Skink
Emoia cyanura	White-bellied Copper-striped Skink
Emoia slevini	Mariana Skink
Lipinia noctua	Pacific Moth Skink
Varanus "indicus"	Pacific Monitor

SNAKES (SQUAMATA: SERPENTES)

Ramphotyphlops braminus	Brahminy Blindsnake
Boiga irregularis	Brown Treesnake

TURTLES (TESTUDINES)

Chelonia mydas	Green Seaturtle
Chelydra serpentina	Snapping Turtle
Eretmochelys imbricata	Hawksbill Seaturtle
Lepidochelys olivacea	Olive Ridley Seaturtle
Pelodiscus sinensis	Chinese Softshell
Trachemys scripta	Red-eared Slider

CROCODILES (CROCODILIA)   NONE

Wake Island

FROGS (ANURA)   NONE

LIZARDS (SQUAMATA)

Gehyra insulensis	Pacific Stump-toed Gecko
Hemidactylus frenatus	Common House Gecko
Cryptoblepharus poecilopleurus	Oceania Snake-eyed Skink
Emoia cyanura	White-bellied Copper-striped Skink

SNAKES (SQUAMATA: SERPENTES) NONE

TURTLES (TESTUDINES)
Chelonia mydas   Green Seaturtle

CROCODILES (CROCODILIA)   NONE

Johnston Atoll

FROGS (ANURA)   NONE

LIZARDS (SQUAMATA)
Hemidactylus frenatus   Common House Gecko
Hemidactylus garnotii   Fox Gecko
Lepidodactylus lugubris   Mourning Gecko
Cryptoblepharus poecilopleurus   Oceania Snake-eyed Skink
                                    [likely Extinct]

SNAKES (SQUAMATA: SERPENTES) NONE

TURTLES (TESTUDINES)
Chelonia mydas   Green Seaturtle

CROCODILES (CROCODILIA)   NONE

WEST- AND NORTH-CENTRAL PACIFIC: MICRONESIA

Palau

FROGS (ANURA)
Rhinella marina   Cane Toad
Platymantis pelewensis   Palau Frog

LIZARDS (SQUAMATA)
Gehyra brevipalmata   Palau Gecko
Gehyra insulensis   Pacific Stump-toed Gecko
Gehyra oceanica   Oceania Gecko
Gekko nsp   Palau Ghost Gecko
Hemidactylus frenatus   Common House Gecko
Hemiphyllodactylus ganoklonis   Palau Slender Gecko

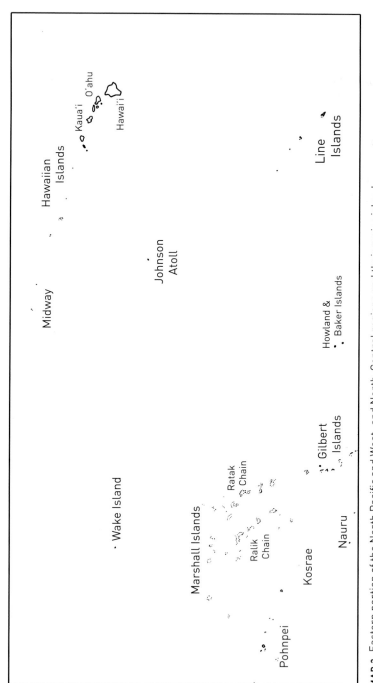

MAP 3. Eastern portion of the North Pacific and West- and North-Central regions and their major island groups.

Lepidodactylus nsp	Palau Flat-tailed Gecko [Iilblau, Ngeruktabel, Ulang]
Lepidodactylus lugubris	Mourning Gecko
Lepidodactylus moestus	Micronesia Flat-tailed Gecko
Lepidodactylus paurolepis	Palau Barred Gecko [Ngerukeuid Islands]
Nactus sp [bisexual]	Palau Slender-toed Gecko [Ulong, Ulebsechel]
Nactus pelagicus	Pacific Slender-toed Gecko
Perochirus nsp	Palau Saw-tailed Gecko [Fana]
Anolis carolinensis	Green Anole
Carlia tutela	Halmahera Brown Skink
Cryptoblepharus rutilus	Palau Snake-eyed Skink
Emoia atrocostata	Seaside Skink
Emoia caeruleocauda	Pacific Blue-tailed Skink
Emoia impar	Dark-bellied Copper-striped Skink
Emoia jakati	Papua Five-striped Skink
Eugongylus nsp	Palau Recluse Skink
Eutropis multicarinata	Micronesia Multi-keeled Sunskink
Lamprolepis smaragdina	Emerald Treeskink
Lipinia leptosoma	Pandanus Moth Skink
Lipinia noctua	Pacific Moth Skink
Sphenomorphus nsp	Palau Large Ground Skink [Ngermalk, Ngeruktabel, Oreor]
Sphenomorphus scutatus	Palau Ground Skink
Varanus "indicus"	Pacific Monitor
Hypsilurus godeffroyi	Papua Angle-headed Lizard

SNAKES (SQUAMATA: SERPENTES)

Ramphotyphlops acuticaudus	Palau Blindsnake
Ramphotyphlops braminus	Brahminy Blindsnake
Candoia superciliosa	Palau Bevel-nosed Boa
Cerberus dunsoni	Palau Dog-faced Mudsnake

Dendrelaphis striolatus	Palau Treesnake
Laticauda colubrina	Yellow-lipped Seakrait
Pelamis platura	Yellow-bellied Seasnake

TURTLES (TESTUDINES)

Mauremys reevesii	Chinese Three-keeled Pondturtle
Chelonia mydas	Green Seaturtle
Eretmochelys imbricata	Hawksbill Seaturtle
Lepidochelys olivacea	Olive Ridley Seaturtle
Dermochelys coriacea	Leatherback Seaturtle

CROCODILES (CROCODILIA)

Crocodylus porosus	Saltwater Crocodile

Micronesia-Yap

FROGS (ANURA)

Rhinella marina	Cane Toad

LIZARDS (SQUAMATA)

Gehyra insulensis	Pacific Stump-toed Gecko
Gehyra oceanica	Oceania Gecko
Hemidactylus frenatus	Common House Gecko
Lepidodactylus lugubris	Mourning Gecko
Lepidodactylus moestus	Micronesia Flat-tailed Gecko
Nactus pelagicus	Pacific Slender-toed Gecko
Perochirus ateles	Micronesia Saw-tailed Gecko
Carlia ailanpalai	Admiralty Brown Skink
Emoia atrocostata	Seaside Skink
Emoia caeruleocauda	Pacific Blue-tailed Skink
Emoia impar	Dark-bellied Copper-striped Skink
Emoia jakati	Papua Five-striped Skink
Emoia nigra	South Pacific Black Skink
Eugongylus albofasciolatus	Barred Recluse Skink

Eutropis multicarinata	Micronesia Multi-keeled Sunskink
Lamprolepis smaragdina	Emerald Treeskink
Lipinia noctua	Pacific Moth Skink
Varanus "indicus"	Pacific Monitor

SNAKES (SQUAMATA: SERPENTES)

Ramphotyphlops braminus	Brahminy Blindsnake
Ramphotyphlops hatmaliyeb	Ulithi Blindsnake

TURTLES (TESTUDINES)

Chelonia mydas	Green Seaturtle
Eretmochelys imbricata	Hawksbill Seaturtle
Lepidochelys olivacea	Olive Ridley Seaturtle
Dermochelys coriacea	Leatherback Seaturtle

CROCODILES (CROCODILIA)   NONE

Micronesia-Chuuk

FROGS (ANURA)   NONE

LIZARDS (SQUAMATA)

Gehyra insulensis	Pacific Stump-toed Gecko
Gehyra oceanica	Oceania Gecko
Hemidactylus frenatus	Common House Gecko
Lepidodactylus lugubris	Mourning Gecko
Lepidodactylus oligoporus	Mortlock Forest Gecko [Mortlock Island]
Perochirus ateles	Micronesia Saw-tailed Gecko
Carlia ailanpalai	Admiralty Brown Skink
Emoia arnoensis	Micronesia Black Skink
Emoia atrocostata	Seaside Skink
Emoia boettgeri	Micronesia Spotted Skink
Emoia caeruleocauda	Pacific Blue-tailed Skink
Emoia cyanura	White-bellied Copper-striped Skink

Emoia impar	Dark-bellied Copper-striped Skink
Emoia jakati	Papua Five-striped Skink
Eugongylus albofasciolatus	Barred Recluse Skink
Lamprolepis smaragdina	Emerald Treeskink
Lipinia noctua	Pacific Moth Skink
Varanus "indicus"	Pacific Monitor

SNAKES (SQUAMATA: SERPENTES)

Pelamis platura	Yellow-bellied Seasnake

TURTLES (TESTUDINES)

Chelonia mydas	Green Seaturtle
Eretmochelys imbricata	Hawksbill Seaturtle
Dermochelys coriacea	Leatherback Seaturtle

CROCODILES (CROCODILIA)   NONE

Micronesia-Pohnpei (includes Kosrae)

FROGS (ANURA)

Rhinella marina	Cane Toad

LIZARDS (SQUAMATA)

Gehyra insulensis	Pacific Stump-toed Gecko
Gehyra oceanica	Oceania Gecko
Hemidactylus frenatus	Common House Gecko
Lepidodactylus lugubris	Mourning Gecko
Lepidodactylus moestus	Micronesia Flat-tailed Gecko
Nactus pelagicus	Pacific Slender-toed Gecko
Perochirus ateles	Micronesia Saw-tailed Gecko
Cryptoblepharus poecilopleurus	Oceania Snake-eyed Skink
Emoia arnoensis	Micronesia Black Skink
Emoia atrocostata	Seaside Skink
Emoia boettgeri	Micronesia Spotted Skink
Emoia caeruleocauda	Pacific Blue-tailed Skink

Emoia cyanura	White-bellied Copper-striped Skink
Emoia impar	Dark-bellied Copper-striped Skink
Emoia jakati	Papua Five-striped Skink
Emoia ponapea	Pohnpei Skink
Eugongylus albofasciolatus	Barred Recluse Skink
Lamprolepis smaragdina	Emerald Treeskink
Lipinia noctua	Pacific Moth Skink
Varanus "indicus"	Pacific Monitor

SNAKES (SQUAMATA: SERPENTES)

Ramphotyphlops adocetus	Ant Atoll Blindsnake
Ramphotyphlops braminus	Brahminy Blindsnake
Laticauda colubrina	Yellow-lipped Seakrait
Pelamis platura	Yellow-bellied Seasnake

TURTLES (TESTUDINES)

Chelonia mydas	Green Seaturtle
Eretmochelys imbricata	Hawksbill Seaturtle

CROCODILES (CROCODILIA)   NONE

Marshall Islands [Ralik and Ratak chains]

FROGS (ANURA)   NONE

LIZARDS (SQUAMATA)

Gehyra oceanica	Oceania Gecko
Hemidactylus frenatus	Common House Gecko
Hemiphyllodactylus typus	Indo-Pacific Slender Gecko
Lepidodactylus lugubris	Mourning Gecko
Lepidodactylus moestus	Micronesia Flat-tailed Gecko
Nactus pelagicus	Pacific Slender-toed Gecko
Perochirus ateles	Micronesia Saw-tailed Gecko
Cryptoblepharus poecilopleurus	Oceania Snake-eyed Skink

Emoia arnoensis	Micronesia Black Skink
Emoia boettgeri	Micronesia Spotted Skink
Emoia caeruleocauda	Pacific Blue-tailed Skink
Emoia cyanura	White-bellied Copper-striped Skink
Emoia impar	Dark-bellied Copper-striped Skink
Emoia jakati	Papua Five-striped Skink
Eugongylus albofasciolatus	Barred Recluse Skink
Lamprolepis smaragdina	Emerald Treeskink
Lipinia noctua	Pacific Moth Skink
Varanus "indicus"	Pacific Monitor

SNAKES (SQUAMATA: SERPENTES)

Ramphotyphlops braminus	Brahminy Blindsnake

TURTLES (TESTUDINES)

Chelonia mydas	Green Seaturtle
Eretmochelys imbricata	Hawksbill Seaturtle
Dermochelys coriacea	Leatherback Seaturtle

CROCODILES (CROCODILIA) — NONE

EQUATORIAL PACIFIC

Kapingamarangi Atoll

FROGS (ANURA) — NONE

LIZARDS (SQUAMATA)

Gehyra insulensis	Pacific Stump-toed Gecko
Gehyra oceanica	Oceania Gecko
Lepidodactylus lugubris	Mourning Gecko
Perochirus ateles	Micronesia Saw-tailed Gecko
Perochirus scutellatus	Giant Saw-tailed Gecko
Emoia impar	Dark-bellied Copper-striped Skink
Lipinia noctua	Pacific Moth Skink

SNAKES (SQUAMATA: SERPENTES) NONE

TURTLES (TESTUDINES)

Chelonia mydas	Green Seaturtle
Eretmochelys imbricata	Hawksbill Seaturtle
Dermochelys coriacea	Leatherback Seaturtle

CROCODILES (CROCODILIA) NONE

Nauru

FROGS (ANURA) NONE

LIZARDS (SQUAMATA)

Gehyra insulensis	Pacific Stump-toed Gecko
Gehyra oceanica	Oceania Gecko
Hemidactylus frenatus	Common House Gecko
Lepidodactylus lugubris	Mourning Gecko
Emoia arnoensis	Micronesia Black Skink
Emoia cyanura	White-bellied Copper-striped Skink

SNAKES (SQUAMATA: SERPENTES)

Ramphotyphlops braminus	Brahminy Blindsnake
Pelamis platura	Yellow-bellied Seasnake

TURTLES (TESTUDINES)

Chelonia mydas	Green Seaturtle
Eretmochelys imbricata	Hawksbill Seaturtle
Dermochelys coriacea	Leatherback Seaturtle

CROCODILES (CROCODILIA) NONE

Western Kiribati (Banaba and "Tarawa" chains) [Gilbert Islands]

FROGS (ANURA)

Rhinella marina	Cane Toad

LIZARDS (SQUAMATA)

Gehyra oceanica	Oceania Gecko
Hemidactylus frenatus	Common House Gecko
Lepidodactylus lugubris	Mourning Gecko
Cryptoblepharus poecilopleurus	Oceania Snake-eyed Skink
Emoia cyanura	White-bellied Copper-striped Skink
Emoia impar	Dark-bellied Copper-striped Skink
Lipinia noctua	Pacific Moth Skink

SNAKES (SQUAMATA: SERPENTES)

Hydrophis belcheri	Questionable record!
Laticauda colubrina	Yellow-lipped Seakrait

TURTLES (TESTUDINES)

Chelonia mydas	Green Seaturtle
Eretmochelys imbricata	Hawksbill Seaturtle
Dermochelys coriacea	Leatherback Seaturtle

CROCODILES (CROCODILIA)   NONE

Howland and Baker islands

FROGS (ANURA)   NONE

LIZARDS (SQUAMATA)

Hemidactylus frenatus	Common House Gecko [suspected, but not confirmed]
Lepidodactylus lugubris	Mourning Gecko
Cryptoblepharus poecilopleurus	Oceania Snake-eyed Skink
Emoia cyanura	White-bellied Copper-striped Skink

SNAKES (SQUAMATA: SERPENTES)   NONE

TURTLES (TESTUDINES)   NONE

One or more species of seaturtles likely; none confirmed

CROCODILES (CROCODILIA)   NONE

Kiribati-northern Line Islands [includes Palmyra Atoll, Jarvis Island, Kirimati]

FROGS (ANURA)	NONE

LIZARDS (SQUAMATA)

Hemidactylus frenatus	Common House Gecko
Lepidodactylus lugubris	Mourning Gecko
Lepidodactylus nsp	Central Pacific Beach Gecko
Emoia cyanura	White-bellied Copper-striped Skink

SNAKES (SQUAMATA: SERPENTES)	NONE

TURTLES (TESTUDINES)

Chelonia mydas	Green Seaturtle
Dermochelys coriacea	Leatherback Seaturtle

CROCODILES (CROCODILIA)	NONE

Tuvalu

FROGS (ANURA)

Rhinella marina	Cane Toad

LIZARDS (SQUAMATA)

Gehyra oceanica	Oceania Gecko
Lepidodactylus lugubris	Mourning Gecko
Lepidodactylus tepukapili	Tuvalu Forest Gecko
Nactus pelagicus	Pacific Slender-toed Gecko
Cryptoblepharus poecilopleurus	Oceania Snake-eyed Skink
Emoia adspersa	Striped Small-scaled Skink
Emoia cyanura	White-bellied Copper-striped Skink
Emoia impar	Dark-bellied Copper-striped Skink
Lipinia noctua	Pacific Moth Skink

SNAKES (SQUAMATA: SERPENTES)

Laticauda colubrina Yellow-lipped Seakrait

Pelamis platura Yellow-bellied Seasnake

TURTLES (TESTUDINES)

Chelonia mydas Green Seaturtle

Eretmochelys imbricata Hawksbill Seaturtle

Dermochelys coriacea Leatherback Seaturtle

CROCODILES (CROCODILIA) NONE

Kiribati-Rawaki

FROGS (ANURA) NONE

LIZARDS (SQUAMATA)

Gehyra insulensis Pacific Stump-toed Gecko

Gehyra oceanica Oceania Gecko

Hemidactylus frenatus Common House Gecko

Lepidodactylus lugubris Mourning Gecko

Lepidodactylus nsp Central Pacific Beach Gecko

Cryptoblepharus poecilopleurus Oceania Snake-eyed Skink

Emoia cyanura White-bellied Copper-striped Skink

SNAKES (SQUAMATA: SERPENTES) NONE

TURTLES (TESTUDINES)

Chelonia mydas Green Seaturtle

Dermochelys coriacea Leatherback Seaturtle

CROCODILES (CROCODILIA) NONE

Tokelau

FROGS (ANURA) NONE

LIZARDS (SQUAMATA)

Gehyra insulensis Pacific Stump-toed Gecko

Gehyra oceanica	Oceania Gecko
Lepidodactylus lugubris	Mourning Gecko
Nactus pelagicus	Pacific Slender-toed Gecko
Cryptoblepharus poecilopleurus	Oceania Snake-eyed Skink
Emoia adspersa	Striped Small-scaled Skink
Emoia cyanura	White-bellied Copper-striped Skink
Emoia impar	Dark-bellied Copper-striped Skink
Emoia nigra	South Pacific Black Skink
Lipinia noctua	Pacific Moth Skink

SNAKES (SQUAMATA: SERPENTES) NONE

TURTLES (TESTUDINES)

Caretta caretta	Loggerhead Seaturtle
Chelonia mydas	Green Seaturtle

CROCODILES (CROCODILIA) NONE

Kiribati–southern Line Islands

FROGS (ANURA) NONE

LIZARDS (SQUAMATA)

Gehyra insulensis	Pacific Stump-toed Gecko
Gehyra oceanica	Oceania Gecko
Hemidactylus frenatus	Common House Gecko [suspected, but not confirmed]
Lepidodactylus lugubris	Mourning Gecko
Lepidodactylus nsp	Central Pacific Beach Gecko
Cryptoblepharus poecilopleurus	Oceania Snake-eyed Skink
Emoia cyanura	White-bellied Copper-striped Skink
Emoia nigra	South Pacific Black Skink
Lipinia noctua	Pacific Moth Skink

SNAKES (SQUAMATA: SERPENTES) NONE

TURTLES (TESTUDINES)

Chelonia mydas — Green Seaturtle

CROCODILES (CROCODILIA) — NONE

SOUTHWEST PACIFIC

Santa Cruz and eastern outer islands – Solomon Islands

FROGS (ANURA) — NONE

LIZARDS (SQUAMATA)

Gehyra insulensis	Pacific Stump-toed Gecko
Gehyra oceanica	Oceania Gecko
Gekko vittatus	Melanesia Ghost Gecko
Hemidactylus frenatus	Common House Gecko
Lepidodactylus guppyi	Solomon Forest Gecko
Lepidodactylus lugubris	Mourning Gecko
Nactus multicarinatus	Melanesia Slender-toed Gecko
Emoia atrocostata	Seaside Skink
Emoia caeruleocauda	Pacific Blue-tailed Skink
Emoia cyanogaster	Green-bellied Vineskink
Emoia cyanura	White-bellied Copper-striped Skink
Emoia nigra	South Pacific Black Skink
Emoia rufilabialis	Red-lipped Striped Skink
Emoia taumakoenis	Taumako Skink
Eugongylus albofasciolatus	Barred Recluse Skink
Lamprolepis smaragdina	Emerald Treeskink
Prasinohaema virens	Green-blooded Vineskink
Sphenomorphus solomonis	Solomon Ground Skink

SNAKES (SQUAMATA: SERPENTES)

Ramphotyphlops depressus	Melanesia Blindsnake
Candoia bibroni	Pacific Treeboa
Candoia paulsoni	Melanesia Bevel-nosed Boa
Boiga irregularis	Brown Treesnake

Dendrelaphis salomonis	Solomon Treesnake
Laticauda colubrina	Yellow-lipped Seakrait
Laticauda laticaudata	Dark-lipped Seakrait

TURTLES (TESTUDINES)

Chelonia mydas	Green Seaturtle
Eretmochelys imbricata	Hawksbill Seaturtle
Lepidochelys olivacea	Olive Ridley Seaturtle
Dermochelys coriacea	Leatherback Seaturtle

CROCODILES (CROCODILIA)

Crocodylus porosus	Saltwater Crocodile

Vanuatu

FROGS (ANURA)

Litoria aurea	Green and Gold Bellfrog

LIZARDS (SQUAMATA)

Gehyra georgpotthasti	Vanuatu Giant Gecko
Gehyra insulensis	Pacific Stump-toed Gecko [Toga Island]
Gehyra oceanica	Oceania Gecko
Gekko vittatus	Melanesia Ghost Gecko
Hemidactylus frenatus	Common House Gecko
Hemidactylus garnotii	Fox Gecko
Hemiphyllodactylus typus	Indo-Pacific Slender Gecko
Lepidodactylus buleli	Vanuatu Ant-nest Gecko [Espiritu Santo]
Lepidodactylus guppyi	Solomon Forest Gecko
Lepidodactylus lugubris	Mourning Gecko
Lepidodactylus vanuatuensis	Vanuatu Forest Gecko
Nactus multicarinatus	Melanesia Slender-toed Gecko
Nactus pelagicus	Pacific Slender-toed Gecko
Perochirus guentheri	Vanuatu Saw-tailed Gecko [Erromango]

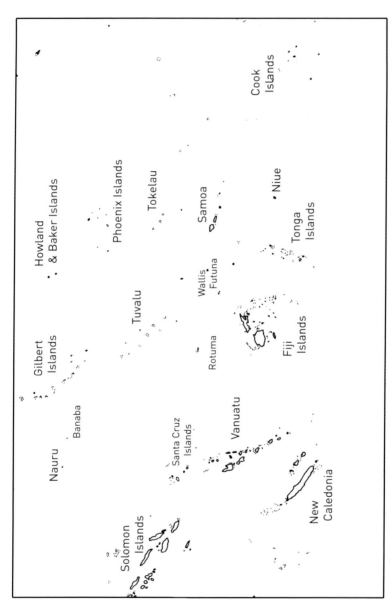

MAP 4. Southwest Pacific region and southern portion of Equatorial region and their major island groups.

Brachylophus bulabula	Fiji Banded Iguana
Caledoniscincus atropunctatus	Speckled Litter Skink
Cryptoblepharus novohebridicus	Vanuatu Snake-eyed Skink
Emoia aneityumensis	Anatom Treeskink
Emoia atrocostata	Seaside Skink
Emoia caeruleocauda	Pacific Blue-tailed Skink
Emoia cyanogaster	Green-bellied Vineskink
Emoia cyanura	White-bellied Copper-striped Skink
Emoia erronan	Erronan Treeskink
Emoia impar	Dark-bellied Copper-striped Skink
Emoia nigra	South Pacific Black Skink
Emoia nigromarginata	Vanuatu Coppery Vineskink
Emoia sanfordi	Toupeed Treeskink
Lipinia noctua	Pacific Moth Skink

SNAKES (SQUAMATA: SERPENTES)

Ramphotyphlops braminus	Brahminy Blindsnake
Candoia bibroni	Pacific Treeboa
Laticauda colubrina	Yellow-lipped Seakrait
Laticauda frontalis	Vanuatu Yellow-lipped Seakrait
Laticauda laticaudata	Dark-lipped Seakrait
Pelamis platura	Yellow-bellied Seasnake

TURTLES (TESTUDINES)

Caretta caretta	Loggerhead Seaturtle
Chelonia mydas	Green Seaturtle
Eretmochelys imbricata	Hawksbill Seaturtle
Lepidochelys olivacea	Olive Ridley Seaturtle
Dermochelys coriacea	Leatherback Seaturtle

CROCODILES (CROCODILIA)

| *Crocodylus porosus* | Saltwater Crocodile |

Rotuma

FROGS (ANURA)   NONE

LIZARDS (SQUAMATA)

Gehyra insulensis   Pacific Stump-toed Gecko

Gehyra oceanica   Oceania Gecko

Lepidodactylus gardineri   Rotuma Forest Gecko

Lepidodactylus lugubris   Mourning Gecko

Nactus pelagicus   Pacific Slender-toed Gecko

Emoia cyanura   White-bellied Copper-striped
                                        Skink

Emoia nigra   South Pacific Black Skink

Emoia oriva   Rotuma Barred Treeskink

Emoia tongana/concolor   Polynesia Slender Treeskink

Lipinia noctua   Pacific Moth Skink

SNAKES (SQUAMATA: SERPENTES)

Candoia bibroni   Pacific Treeboa

TURTLES (TESTUDINES)   NONE

One or more seaturtles likely; none confirmed.

CROCODILES (CROCODILIA)   NONE

Fiji

FROGS (ANURA)

Rhinella marina   Cane Toad

Platymantis vitiana   Fiji Ground Frog

Platymantis vitiensis   Fiji Treefrog

Platymantis nsp   Taveuni Treefrog

LIZARDS (SQUAMATA)

Gehyra insulensis   Pacific Stump-toed Gecko

Gehyra oceanica   Oceania Gecko

Gehyra vorax   Fiji Giant Gecko

Hemidactylus frenatus	Common House Gecko
Hemidactylus garnotii	Fox Gecko
Hemiphyllodactylus typus	Indo-Pacific Slender Gecko
Lepidodactylus lugubris	Mourning Gecko
Lepidodactylus manni	Fiji Forest Gecko
Nactus pelagicus	Pacific Slender-toed Gecko
Brachylophus bulabula	Fiji Banded Iguana
Brachylophus fasciatus	Lau Banded Iguana
Brachylophus vitiensis	Fiji Crested Iguana
Cryptoblepharus eximius	Fiji Snake-eyed Skink
Emoia caeruleocauda	Pacific Blue-tailed Skink
Emoia campbelli	Vitilevu Mountain Treeskink
Emoia concolor	Fiji Slender Treeskink
Emoia cyanura	White-bellied Copper-striped Skink
Emoia impar	Dark-bellied Copper-striped Skink
Emoia mokosariniveikau	Vanualevu Slender Treeskink
Emoia nigra	South Pacific Black Skink
Emoia parkeri	Fiji Copper-headed Skink
Emoia trossula	Fiji Barred Treeskink
Leiolopisma alazon	Ono-i-Lau Ground Skink
Lipinia noctua	Pacific Moth Skink

SNAKES (SQUAMATA: SERPENTES)

Ramphotyphlops nsp	Taveuni Blindsnake
Ramphotyphlops braminus	Brahminy Blindsnake
Candoia bibroni	Pacific Treeboa
Hydrophis coggeri	Pacific Yellow-banded Seasnake
Laticauda colubrina	Yellow-lipped Seakrait
Ogmodon vitianus	Fiji Bola
Pelamis platura	Yellow-bellied Seasnake

TURTLES (TESTUDINES)

Caretta caretta	Loggerhead Seaturtle
Chelonia mydas	Green Seaturtle
Eretmochelys imbricata	Hawksbill Seaturtle
Lepidochelys olivacea	Olive Ridley Seaturtle
Dermochelys coriacea	Leatherback Seaturtle

CROCODILES (CROCODILIA)   NONE

Tonga

FROGS (ANURA)   NONE

LIZARDS (SQUAMATA)

Gehyra insulensis	Pacific Stump-toed Gecko
Gehyra oceanica	Oceania Gecko
Hemidactylus frenatus	Common House Gecko
Hemidactylus garnotii	Fox Gecko
Hemiphyllodactylus typus	Indo-Pacific Slender Gecko
Lepidodactylus euaensis	'Eua Forest Gecko
Lepidodactylus lugubris	Mourning Gecko
Nactus pelagicus	Pacific Slender-toed Gecko
Brachylophus fasciatus	Lau Banded Iguana
Cryptoblepharus poecilopleurus	Oceania Snake-eyed Skink
Emoia adspersa	Striped Small-scaled Skink
Emoia cyanura	White-bellied Copper-striped Skink
Emoia impar	Dark-bellied Copper-striped Skink
Emoia lawesii	Olive Small-scaled Skink
Emoia nigra	South Pacific Black Skink
Emoia tongana	Polynesia Slender Treeskink
Emoia mokolahi	Tonga Robust Treeskink
Lipinia noctua	Pacific Moth Skink
Tachygyia microlepis	Small-scaled Giant Skink [Extinct]

SNAKES (SQUAMATA: SERPENTES)

Candoia bibroni	Pacific Treeboa
Laticauda colubrina	Yellow-lipped Seakrait
Laticauda laticaudata	Dark-lipped Seakrait [likely misidentification]
Laticauda schistorhyncha	Central Pacific Seakrait

TURTLES (TESTUDINES)

Caretta caretta	Loggerhead Seaturtle
Chelonia mydas	Green Seaturtle
Eretmochelys imbricata	Hawksbill Seaturtle
Lepidochelys olivacea	Olive Ridley Seaturtle
Dermochelys coriacea	Leatherback Seaturtle

CROCODILES (CROCODILIA)   NONE

Wallis and Futuna

FROGS (ANURA)   NONE

LIZARDS (SQUAMATA)

Gehyra oceanica	Oceania Gecko
Hemidactylus frenatus	Common House Gecko
Lepidodactylus lugubris	Mourning Gecko
Nactus pelagicus	Pacific Slender-toed Gecko
Cryptoblepharus poecilopleurus	Oceania Snake-eyed Skink
Emoia adspersa	Striped Small-scaled Skink
Emoia cyanura	White-bellied Copper-striped Skink
Emoia impar	Dark-bellied Copper-striped Skink
Emoia nigra	South Pacific Black Skink
Emoia oriva	Rotuma Barred Treeskink
Emoia tongana	Polynesia Slender Treeskink

SNAKES (SQUAMATA: SERPENTES)

Candoia bibroni	Pacific Treeboa

TURTLES (TESTUDINES)

Chelonia mydas	Green Seaturtle
Eretmochelys imbricata	Hawksbill Seaturtle

CROCODILES (CROCODILIA)   NONE

Samoa

FROGS (ANURA)

Rhinella marina	Cane Toad

LIZARDS (SQUAMATA)

Gehyra insulensis	Pacific Stump-toed Gecko
Gehyra oceanica	Oceania Gecko
Hemidactylus frenatus	Common House Gecko
Hemidactylus garnotii	Fox Gecko
Hemiphyllodactylus typus	Indo-Pacific Slender Gecko
Lepidodactylus lugubris	Mourning Gecko
Nactus pelagicus	Pacific Slender-toed Gecko
Cryptoblepharus poecilopleurus	Oceania Snake-eyed Skink
Emoia adspersa	Striped Small-scaled Skink
Emoia cyanura	White-bellied Copper-striped Skink
Emoia impar	Dark-bellied Copper-striped Skink
Emoia lawesii	Olive Small-scaled Skink
Emoia nigra	South Pacific Black Skink
Emoia tongana	Polynesia Slender Treeskink
Emoia samoensis	Pacific Robust Treeskink
Lipinia noctua	Pacific Moth Skink

SNAKES (SQUAMATA: SERPENTES)

Ramphotyphlops braminus	Brahminy Blindsnake
Candoia bibroni	Pacific Treeboa
Laticauda colubrina	Yellow-lipped Seakrait
Laticauda schistorhyncha	Central Pacific Seakrait

TURTLES (TESTUDINES)

Caretta caretta	Loggerhead Seaturtle
Chelonia mydas	Green Seaturtle
Eretmochelys imbricata	Hawksbill Seaturtle
Dermochelys coriacea	Leatherback Seaturtle

CROCODILES (CROCODILIA)   NONE

Niue

FROGS (ANURA)   NONE

LIZARDS (SQUAMATA)

Gehyra oceanica	Oceania Gecko
Lepidodactylus lugubris	Mourning Gecko
Nactus pelagicus	Pacific Slender-toed Gecko
Cryptoblepharus poecilopleurus	Oceania Snake-eyed Skink
Emoia cyanura	White-bellied Copper-striped Skink
Emoia impar	Dark-bellied Copper-striped Skink
Emoia lawesii	Olive Small-scaled Skink
Lipinia noctua	Pacific Moth Skink

SNAKES (SQUAMATA: SERPENTES)

Laticauda schistorhyncha	Central Pacific Seakrait

TURTLES (TESTUDINES)

Chelonia mydas	Green Seaturtle

CROCODILES (CROCODILIA)   NONE

SOUTH-CENTRAL PACIFIC: POLYNESIA

Cook Islands

FROGS (ANURA)   NONE

LIZARDS (SQUAMATA)

Gehyra insulensis	Pacific Stump-toed Gecko

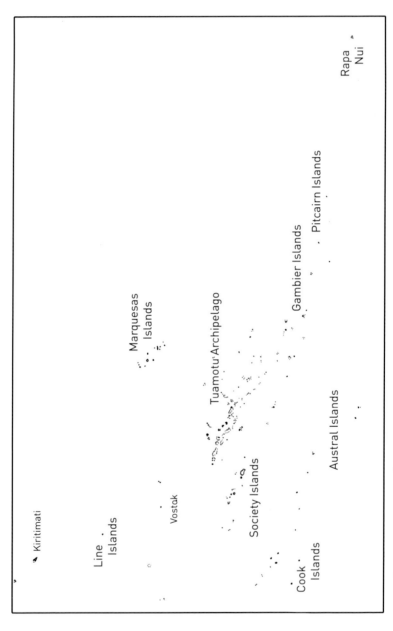

MAP 5. South-Central region (core of Polynesia) and its major island groups.

Gehyra oceanica	Oceania Gecko
Hemidactylus frenatus	Common House Gecko
Hemidactylus garnotii	Fox Gecko
Hemiphyllodactylus typus	Indo-Pacific Slender Gecko
Lepidodactylus lugubris	Mourning Gecko
Lepidodactylus nsp	Central Pacific Beach Gecko
Nactus pelagicus	Pacific Slender-toed Gecko [Nassau]
Cryptoblepharus poecilopleurus	Oceania Snake-eyed Skink
Emoia adspersa	Striped Small-scaled Skink
Emoia cyanura	White-bellied Copper-striped Skink
Emoia impar	Dark-bellied Copper-striped Skink
Emoia tuitarere	Rarotonga Treeskink [Rarotonga]
Lipinia noctua	Pacific Moth Skink

SNAKES (SQUAMATA: SERPENTES)

Pelamis platura	Yellow-bellied Seasnake

TURTLES (TESTUDINES)

Chelodina longicollis	Eastern Long-necked Turtle
Caretta caretta	Loggerhead Seaturtle
Chelonia mydas	Green Seaturtle
Eretmochelys imbricata	Hawksbill Seaturtle
Dermochelys coriacea	Leatherback Seaturtle

CROCODILES (CROCODILIA)   NONE

Austral and Society Islands

FROGS (ANURA)   NONE

LIZARDS (SQUAMATA)

Gehyra insulensis	Pacific Stump-toed Gecko
Gehyra oceanica	Oceania Gecko

Hemidactylus frenatus	Common House Gecko
Hemidactylus garnotii	Fox Gecko
Hemiphyllodactylus typus	Indo-Pacific Slender Gecko
Lepidodactylus lugubris	Mourning Gecko
Lepidodactylus nsp	Central Pacific Beach Gecko
Nactus pelagicus	Pacific Slender-toed Gecko [Tahiti]
Phelsuma laticauda	Golddust Daygecko
Cryptoblepharus poecilopleurus	Oceania Snake-eyed Skink
Emoia cyanura	White-bellied Copper-striped Skink
Emoia impar	Dark-bellied Copper-striped Skink
Lipinia noctua	Pacific Moth Skink

SNAKES (SQUAMATA: SERPENTES)

Pelamis platura	Yellow-bellied Seasnake

TURTLES (TESTUDINES)

Chelonia mydas	Green Seaturtle
Trachemys scripta	Red-eared Slider

CROCODILES (CROCODILIA)   NONE

Tuamotu Archipelago and Gambier Island

FROGS (ANURA)   NONE

LIZARDS (SQUAMATA)

Gehyra insulensis	Pacific Stump-toed Gecko
Gehyra oceanica	Oceania Gecko
Hemidactylus frenatus	Common House Gecko
Hemidactylus garnotii	Fox Gecko
Lepidodactylus lugubris	Mourning Gecko
Nactus pelagicus	Pacific Slender-toed Gecko [Aki Aki]
Cryptoblepharus poecilopleurus	Oceania Snake-eyed Skink

Emoia cyanura	White-bellied Copper-striped Skink
Emoia impar	Dark-bellied Copper-striped Skink
Lipinia noctua	Pacific Moth Skink
SNAKES (SQUAMATA: SERPENTES)	
Pelamis platura	Yellow-bellied Seasnake
TURTLES (TESTUDINES)	
Chelonia mydas	Green Seaturtle
CROCODILES (CROCODILIA)	NONE

Marquesas Islands

FROGS (ANURA)	NONE
LIZARDS (SQUAMATA)	
Gehyra insulensis	Pacific Stump-toed Gecko
Gehyra oceanica	Oceania Gecko
Hemidactylus frenatus	Common House Gecko
Hemidactylus garnotii	Fox Gecko
Hemiphyllodactylus typus	Indo-Pacific Slender Gecko
Lepidodactylus lugubris	Mourning Gecko
Lepidodactylus nsp	Central Pacific Beach Gecko
Cryptoblepharus poecilopleurus	Oceania Snake-eyed Skink
Emoia cyanura	White-bellied Copper-striped Skink
Emoia impar	Dark-bellied Copper-striped Skink
Lipinia noctua	Pacific Moth Skink
SNAKES (SQUAMATA: SERPENTES)	NONE
TURTLES (TESTUDINES)	
Chelonia mydas	Green Seaturtle
CROCODILES (CROCODILIA)	NONE

Pitcairn Islands

FROGS (ANURA)   NONE

LIZARDS (SQUAMATA)
Gehyra insulensis   Pacific Stump-toed Gecko
Hemiphyllodactylus typus   Indo-Pacific Slender Gecko
Lepidodactylus lugubris   Mourning Gecko
Cryptoblepharus poecilopleurus   Oceania Snake-eyed Skink
Emoia cyanura   White-bellied Copper-striped Skink

Lipinia noctua   Pacific Moth Skink

SNAKES (SQUAMATA: SERPENTES) NONE

TURTLES (TESTUDINES)   NONE

CROCODILES (CROCODILIA)   NONE

EASTERN PACIFIC
Islas Revillagigedo

FROGS (ANURA)   NONE

LIZARDS (SQUAMATA)
Hemidactylus frenatus   Common House Gecko
Urosaurus auriculatus   Socorro Treelizard
Urosaurus clarionensis   Clarion Treelizard

SNAKES (SQUAMATA: SERPENTES)
Masticophis anthonyi   Clarion Racer

TURTLES (TESTUDINES)
Chelonia mydas   Green Seaturtle
Eretmochelys imbricata   Hawksbill Seaturtle
Lepidochelys olivacea   Olive Ridley Seaturtle
Dermochelys coriacea   Leatherback Seaturtle

CROCODILES (CROCODILIA)   NONE

Isla del Coco

FROGS (ANURA)	NONE

LIZARDS (SQUAMATA)

Sphaerodactylus pacificus	Cocos Pygmy Gecko
Anolis townsendi	Cocos Anole

SNAKES (SQUAMATA: SERPENTES)

Pelamis platura	Yellow-bellied Seasnake

TURTLES (TESTUDINES)

Chelonia mydas	Green Seaturtle
Eretmochelys imbricata	Hawksbill Seaturtle
Lepidochelys olivacea	Olive Ridley Seaturtle

CROCODILES (CROCODILIA)	NONE

Isla Malpelo

FROGS (ANURA)	NONE

LIZARDS (SQUAMATA)

Diploglossus millepunctatus	Malpelo Galliwasp
Lepidodactylus lugubris	Mourning Gecko
Phyllodactylus transversalis	Malpelo Leaf-toed Gecko
Anolis agassizii	Malpelo Anole

SNAKES (SQUAMATA: SERPENTES)	NONE
TURTLES (TESTUDINES)	Likely several seaturtle species; none verified.
CROCODILES (CROCODILIA)	NONE

Clipperton Island

FROGS (ANURA)	NONE

LIZARDS (SQUAMATA)

Gehyra insulensis	Pacific Stump-toed Gecko
Emoia cyanura	White-bellied Copper-striped Skink

SNAKES (SQUAMATA: SERPENTES)

Pelamis platura — Yellow-bellied Seasnake

TURTLES (TESTUDINES)

Lepidochelys olivacea — Olive Ridley Seaturtle

CROCODILES (CROCODILIA) — NONE

Galápagos Archipelago

FROGS (ANURA)

Scinax quinquefasciatus — Five-lined Snouted Treefrog

LIZARDS (SQUAMATA)

Gonatodes caudiscutatus	Shield-headed Gecko
Hemidactylus frenatus	Common House Gecko
Lepidodactylus lugubris	Mourning Gecko
Phyllodactylus barringtonensis	Santa Fé Leaf-toed Gecko
Phyllodactylus baurii	Pinta Leaf-toed Gecko
Phyllodactylus darwini	Darwin's Leaf-toed Gecko
Phyllodactylus galapagensis	Galápagos Leaf-toed Gecko
Phyllodactylus gilberti	Wolf Leaf-toed Gecko
Phyllodactylus leei	San Cristóbal Leaf-toed Gecko
Phyllodactylus reissi	Guayaquil Leaf-toed Gecko
Amblyrhynchus cristatus	Marine Iguana
Conolophus marthae	Pink Land Iguana
Conolophus pallidus	Santa Fé Land Iguana
Conolophus subcristatus	Galápagos Land Iguana
Microlophus albemarlensis	Galápagos Lava Lizard
Microlophus bivittatus	San Cristóbal Lava Lizard
Microlophus delanonis	Española Lava Lizard
Microlophus duncanensis	Pinzón Lava Lizard
Microlophus grayii	Floreana Lava Lizard
Microlophus habelii	Marchena Lava Lizard
Microlophus pacificus	Pinta Lava Lizard

SNAKES (SQUAMATA: SERPENTES)

Alsophis biseralis	Galápagos Racer
Antillophis slevini	Galápagos Banded Snake
Antillophis steindachneri	Galápagos Striped Snake
Philodryas hoodensis	Española Racer
Pelamis platura	Yellow-bellied Seasnake

TURTLES (TESTUDINES)

Chelonia mydas	Green Seaturtle
Eretmochelys imbricata	Hawksbill Seaturtle
Lepidochelys olivacea	Olive Ridley Seaturtle
Dermochelys coriacea	Leatherback Seaturtle
Chelonoidis abingdonii	Pinta Giant Tortoise [Extinct]
Chelonoidis becki	Wolf Giant Tortoise
Chelonoidis chathamensis	San Cristóbal Giant Tortoise
Chelonoidis darwini	Santiago Giant Tortoise
Chelonoidis ephippium	Pinzón Giant Tortoise
Chelonoidis guentheri	Sierra Negra Giant Tortoise
Chelonoidis hoodensis	Española Giant Tortoise
Chelonoidis microphyes	Darwin Giant Tortoise
Chelonoidis porteri	La Caseta Giant Tortoise
Chelonoidis vandenburghi	Alcedo Giant Tortoise
Chelonoidis vicina	Cerro Azul Giant Tortoise
Chelonoidis sp	Cerro Fatal Giant Tortoise

CROCODILES (CROCODILIA)   NONE

Rapa Nui [Easter Island, Isla de Pascua]

FROGS (ANURA)   NONE

LIZARDS (SQUAMATA)

Gehyra insulensis	Pacific Stump-toed Gecko
Lepidodactylus lugubris	Mourning Gecko
Cryptoblepharus poecilopleurus	Oceania Snake-eyed Skink

SNAKES (SQUAMATA: SERPENTES)

Pelamis platura   Yellow-bellied Seasnake

TURTLES (TESTUDINES)

Chelonia mydas   Green Seaturtle

Eretmochelys imbricata   Hawksbill Seaturtle

CROCODILES (Crocodilia)   NONE

Islas Sala y Gómez

FROGS (ANURA)   NONE

LIZARDS (SQUAMATA)   NONE

SNAKES (SQUAMATA: SERPENTES)   NONE

TURTLES (TESTUDINES)   Likely 1 or 2 seaturtle species; none verified.

CROCODILES (CROCODILIA)   NONE

SELECTED REFERENCES

Bauer, A. M. and R. A. Sadlier. 2000. *The Herpetofauna of New Caledonia.* Ithaca, NY: Society for the Study of Amphibians and Reptiles.

Crombie, R. I., and G. K. Pregill. 1999. "A checklist of the herpetofauna of the Palau Islands (Republic of Belau), Oceania." *Herpetological Monograph* 13: 29–80.

Fitter, J., D. Fitter, and D. Hosking. 2000. *Wildlife of the Galápagos.* Princeton: Princeton University Press.

Frost, Darrel R. 2011. *Amphibian Species of the World: An Online Reference.* Version 5.5 (31 January, 2011). Electronic Database accessible at http://research.amnh.org/vz/herpetology/amphibia/ American Museum of Natural History, New York, USA.

Gill, B. and T. Whitaker. 1996. *New Zealand Frogs & Reptiles.* Auckland: Bateman Fieldguides.

Kraus, F. 2009. *Alien Reptiles and Amphibians: A Scientific Compendium and Analysis.* New York: Springer Science.

Ineich, I. and C.P. Blanc. 1988. "Distribution des reptiles terrestres en Polynesie orientale." *Atoll Research Bulletin* 318: 1–75.

Lever, C. 2003. *Naturalized Reptiles and Amphibians of the World.* Oxford: Oxford University Press.

McCoy, M. 2006. *Reptiles of the Solomon Islands.* Sofia: Pensoft Publishers.

Morrison, C. 2003. *A Fieldguide to the Herpetofauna of Fiji.* Suva, Fiji: Institute of Applied Sciences, University of the South Pacific.

Swash, A. and R. Still. 2005. *Birds, Mammals, and Reptiles of the Galápagos Islands: An Identification Guide.* 2nd ed. New Haven: Yale University Press.

Uetz, P., ed. *The Reptile Database.* Accessible at http://www.reptile-database .org; copyright © 1995–2011.

Vogt, S. R. andL.L. Williams. 2004. *Common Flora and Fauna of the Mariana Islands.* Saipan: privately printed.

Zug, G. R. 1991. "Lizards of Fiji: Natural history and systematics." *Bishop Museum Bulletin in Zoology* 21:1–136.

Recognizing Species

Each animal group has a set of traits that biologists use to identify the different species (taxa). Size is one of the major features used for identification, although it is confounded by growth. Size data are for sexually mature individuals (adults) herein, unless stated otherwise. In the general descriptions, "size" refers to snout-vent length (SVL) for amphibians, lizards, and snakes. Total length (TotL = SVL + tail length) is given for snakes when SVL is not readily available. Shell length (straight-line carapace length, or SCL) is the standard for turtles. Because size differs among the groups, the descriptors "small," "medium," or "midsize," "large," "very large," and "giant" refer to different size classes. For *frogs*, small is < 36 mm SVL, midsize is 36–72 mm SVL, and large is > 72 mm SVL; for *lizards*, small is < 40 mm SVL, midsize is 40–100 mm SVL, large is 101–200 mm SVL, very large is 201–1,000 mm SVL, and giant is > 1 m SVL; for *snakes*, small is < 300 mm TotL, midsize is 300–800 mm TotL, large is 801–1,500 mm TotL, very large is > 1.5 m TotL, and giant is > 3 m TotL; and for *turtles*, small is < 150 mm SCL, midsize is 150–400 mm SCL, large is 401–800 mm SCL, very large is 81–120 cm SCL, and giant is > 120 cm SCL.

Technical terminology is kept to a minimum in the descriptions, but some more explicit anatomical terms are used to describe position and body part (Fig. 1). "Dorsal" (adjective) and "dorsum" (noun) refer to

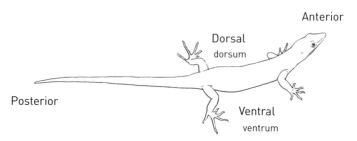

FIGURE 1. Postural and directional terms for vertebrate morphology.

the back of the animal from the top of the snout across the back to the upper-side of tail. "Lateral" refers to the side of an individual—either of the whole animal or just the head. "Ventral" and "venter" refer to the underside from tip of chin to tip of tail; "chin," "throat," "chest," and "abdomen" (belly) are specific areas of the venter. "Anterior" references both direction and region—that is, the front end of the animal or just its head, whereas "posterior" refers to the area behind the head or the rear end. These five terms are combined regularly to enhance the description of position. "Dorsolateral" is perhaps the most frequent combination because of the presence of dorsolateral folds on the bodies of some frogs and contrasting light or dark stripes on the border at the back and sides of lizards. Two other positional terms are "proximal" (toward or near the midline of the body) and "distal" (away from or distant from the midline).

Some technical anatomical terms cannot be avoided because no common term exists. "Vent" appears in "snout-vent length," for example. Everyone knows what "snout" means, but what does "vent" mean? The vent is the opening of the cloaca to the outside. The posterior body opening in amphibians and reptiles is not the anus. The anus, which is the exit of the digestive tract, is within the cloaca—the common depository space into which the digestive, excretory, and reproductive ducts empty and that leads to the outside through the vent. The "loreal region" is the triangular area between the naris (nostril) and the eye; "face" is used occasionally for this general region and commonly includes the eyes and lips. Other features, such "temporal region" (jowls), "canthus" (edge between the top of the snout and the loreal region), "ear-opening" (usually inset in lizards), and "tympanum" (eardrum, which is on the surface in frogs and turtles) are used less frequently and are usually self-explanatory.

Just as the anatomical terminology is intentionally less technical, the descriptions in the accounts are intended for non-herpetologists and for users with an animal in hand or in close observation. Accounts begin with a comment on general size, followed by a description of surface texture and coloration (both color and pattern). Subsequent statements emphasize traits that fine-tune the identification to a single species. These latter statements are more "herpetological" in detail, such as the amount of toe-webbing and the shape of the digit tip in frogs, or the number of scale rows in lizards and snakes. These traits permit a more precise identification.

Coloration includes both an individual's color and the arrangement or pattern of contrasting color against a background color. "Background color" commonly is abbreviated to "ground color" by herpetologists and that usage is retained here. A variety of markings of different shapes and intensity occurs on frogs and reptiles. "Spots" refer to nearly circular marks ranging in size from tiny dots on a single scale to larger dots that may cover several scales; "blotches" are irregular circular to rectangular marks. A "reticulate pattern" is a netlike pattern of fine lines, whereas a "mottled pattern" consists of larger and irregularly connected spots and blotches. "Stripes" are lines of contrasting color of various widths, either longitudinally or diagonally oriented on the body. When the lines or marks are transverse (crossing and more or less perpendicular to the longitudinal body axis), they become "bars." "Bands" are less rigorously defined, although the term commonly refers to a bar that completely or nearly encircles the body. All these marks can be sharp (with straight or smooth edges) or irregularly edged. Patterns can fade with an individual's increasing age; they are then labeled "diffuse," in contrast to a bright or bold pattern of strongly contrasting colors and distinct marks.

Amphibian skin is always moist because it is a major respiratory surface. The skin of frogs can range from smooth with no or few surface protrusions to very rough or irregular. "Warts" and "warty" are terms that are regularly applied to a toad's skin, and sometimes to other frogs as well, but "tubercles" also applies to rounded elevations of various sizes on the skin. Frog skin also can bear "rugae" (the singular is "ruga"), which are elongated narrow glandular ridges or folds. Many frogs have a long ridge of glandular tissue on each side at the juncture between the back (dorsum) and side (lateral); these ridges are the dorsolateral folds and a useful identification feature. Amphibian skin can be both tuberculate and rugose or just a single state.

Reptile skin is dry and keratinous. Keratin is the hard compound that makes up hair, fingernails, claws, and scales. It is largely impermeable and thus resists water loss. Gecko skin is typically granular (with small, rounded scales) and/or tuberculate; "tubercles" are round scales of various sizes elevated above the surrounding granular scales. Reptile scales range from small and granular like those of geckos to the large scutes covering the shells of tortoises. Scale size commonly varies on an individual. For example, most geckos have small granular, juxtaposed (abutting) scales on the dorsal and lateral surface of the trunk; the scales gradually enlarge ventrolaterally, becoming "typical" flattened, imbricate scales on the venter. These typical scales can have a smooth or keeled (ridged) surface. A keeled scale can have a single keel (unicarinate) or multiple keels (bicarinate and tricarinate are the most frequent multiple-keeled patterns).

In addition to scale size, the number of scales is often a diagnostic feature of a species or group of species. For lizards, three counts are common: midbody scale rows, dorsal scale rows, and fourth-toe lamellae. The "midbody count" refers to the number of scales rows encircling the body at mid-trunk. The "dorsal count" is the number of scale rows from the last enlarged head scales (usually the parietals) to the end of the trunk (transverse line from the rear edge of hind limb insertions). The number of "fourth-toe lamellae" refers to the transversely enlarged scales on the underside of the fourth toe of hindfoot.

The major scale counts for snakes are the number of dorsal trunk scale rows behind the neck (anterior), at midbody, and in front of the vent, and the number of ventral scale rows. For snake descriptions, I report only the number of dorsal scale rows at midbody for most species. The number of ventral scale rows is counted from the first ventral scale twice as wide as long and that is bordered on both sides by the first row of dorsal scales, up to but not including the cloacal scale immediately anterior to the vent.

The preceding features represent a simplified set of the characters used by herpetologists in the identification of amphibian and reptiles species. Although simplified, this set of descriptors in association with the figures and identification plates should permit the user of this guide to identify a Pacific "herp" when in hand.

Identification Plates

PLATE 1 | Pacific Forest Frogs

A. Palau Frog
(Platymantis pelewensis)
PAGE 58
Palau, most islands. Adults
34–54 mm SVL. Face with
dark mask; lips light and
dark barred; small circular
pads on hindfeet digit tips.

B. Palau Frog
(Platymantis pelewensis)
PAGE 58

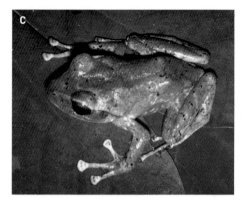

C. Fiji Treefrog
(Platymantis vitiensis)
PAGE 60
Fiji, basaltic islands. Adults
32–60 mm SVL. Face with
dark canthal stripe; large
triangular pads on forefeet
digit tips.

D. Fiji Treefrog
(Platymantis vitiensis)
PAGE 60

E. Fiji Ground Frog
(Platymantis vitiana)
PAGE 59
Fiji, basaltic islands. Adults
40–110 mm SVL. Light
colored spot behind tympanum; small circular pads on
hindfeet digit tips.

F. Fiji Ground Frog
(Platymantis vitiana)
PAGE 59

PLATE 2 | Treefrogs: Aliens

A. Green and Gold Bellfrog
(Litoria aurea) PAGE 69
Vanuatu, Efate, Espiritu Santo.
Adults 57–108 mm SVL. White
to yellow dorsolateral stripe from
snout to posterior trunk.

B. Green and Gold Bellfrog
(Litoria aurea) PAGE 69

C. Eastern Dwarf Treefrog
(Litoria fallax) PAGE 70
Guam. Adults 22–32 mm SVL.
White lip stripe ends at shoulder.

D. Eastern Dwarf Treefrog
(Litoria fallax)   PAGE 70

E. Five-lined Snouted Treefrog
(Scinax quinquefasciatus)
PAGE 70
Galápagos, Santa Cruz. Adults 33–38 mm SVL. Pair of brown stripes on each side of trunk; large pads on all digits of fore- and hindfeet.

F. Taiwan Whipping Frog
(Polypedates braueri)
PAGE 77
Guam. Adults 45–75 mm SVL. Dark brown mask above light colored lips; large pads on all digits of fore- and hindfeet.

PLATE 3 | Fork-tongued and Water Frogs: Aliens

A. Crab-eating Frog
(*Fejervarya cancrivora*)   PAGE 62
Guam. Adults 60–103 mm SVL.
Dorsal trunk skin largely smooth
with a few, short rugae; large
vomerine ridge on roof of mouth.

B. Paddy Frog
(*Fejervarya "limnocharis"*)
   PAGE 63
Guam. Adults 38–76 mm SVL.
Dorsal trunk skin with numerous
rugae of various lengths; small
vomerine ridge on roof of mouth.

C. Brown and Tan Amoy Frog
(*Hylarana guentheri*)   PAGE 74
Guam. Adults 64–80 mm SVL.
Dorsal trunk skin smooth and
prominent dorsolateral glandular
folds.

D. Japanese Wrinkled Frog
(Glandirana rugosa)
PAGE 73

Hawaiian Islands: most islands. Adults 37–53 mm SVL. Dorsal trunk skin strongly rugose; venter dusky brown.

E. American Bullfrog
(Lithobates catesbeianus)
PAGE 75

Hawaiian Islands: most islands. Adults 110–180 mm SVL. Dorsal trunk skin smooth; no dorsolateral folds; venter with dark mottled pattern.

PLATE 4 | Miscellaneous Frogs and Toads: Aliens

A. Cane Toad
(Rhinella marina) female   PAGE 56
Pacific, widespread. Adults 75–180 mm SVL. Large, oblong parotoid gland on side of head in rear; numerous warts on trunk; all digit tips blunt.

B. Cane Toad
(Rhinella marina) male   PAGE 56

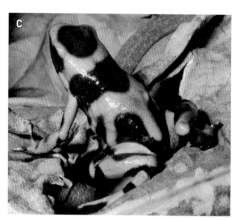

C. Green and Black Poison-dart Frog
(Dendrobates auratus)   PAGE 61
Hawaiian Islands: Maui, Oʻahu. Adults 20–32 mm SVL. Shiny black ground color; all digit tips with moderately large pads.

D. Coquí
(Eleutherodactylus coqui)
PAGE 65
Hawaiian Islands: Hawai'i.
Adults 29–58 mm SVL.
Dorsal skin finely tuber-
culate or rugose; all digits
with large pads.

E. Greenhouse Frog
*(Eleutherodactylus
planirostris)* PAGE 67
Hawaiian Islands: Hawai'i,
O'ahu, and Guam. Adults
15–26 mm SVL. Dorsal skin
coarsely tuberculate; all
digits with large pads.

F. Marbled Pygmy Frog
(Microhyla pulchra)
PAGE 72
Guam. Adults 24–34 mm
SVL. Narrow, pointed head;
all digits end bluntly.

PLATE 5 | Stump-toed Geckos: *Gehyra*

A. Palau Gecko *(Gehyra brevipalmata)* PAGE 80
Palau, most islands. Adults 62–89 mm SVL. Tail subcylindrical without distinct segmentation and covered with smooth granular scales on top and sides; distinct digital pads with 11–13 subdigital lamellae on fourth digit of hindfoot.

B. Pacific Stump-toed Gecko *(Gehyra insulensis)* PAGE 82
Central and western Pacific. Adults 36–50 mm SVL. Tail flattened cylindrical without distinct segmentation and with a ventrolateral border of small elongate scales; distinct digital pads with 8–9 subdigital lamellae on fourth digit of hindfoot.

C. Oceania Gecko *(Gehyra oceanica)* PAGE 84
Central and western Pacific. Adults 59–84 mm SVL. Thick subcylindrical tail with distinct segmentation, each segment with 1–2 spines at ventrolateral edge of each segment; large, circular digital pads with 12–16 subdigital lamellae on fourth digit of hindfoot.

D. Oceania Gecko *(Gehyra oceanica)* PAGE 84

E. Fiji Giant Gecko *(Gehyra vorax)* PAGE 85
Fiji, perhaps Tonga. Adults 90–150 mm SVL. Thick subcylindrical tail with distinct segmentation, encircled by alternating bands of granular and larger smooth scales; large, circular digital pads with 23–34 subdigital lamellae on fourth digit of hindfoot.

PLATE 6 | Geckos: Large Species

A. Melanesia Ghost Gecko *(Gekko vittatus)* PAGE 88
Solomon Islands, widespread. Adults 84–124 mm SVL. Tail subcylindrical, distinctly segmented, each segment with 2–3 whorls of pointed
scales; digital pad with 18–26 subdigital lamellae on fourth digit of
hindfoot.

B. Palau Ghost Gecko *(Gekko nsp)* PAGE 89
Palau, most islands. Adults 102–121 mm SVL. Tail subcylindrical,
distinctly segmented, each segment with 2–3 whorls of pointed scales;
digital pad with 20–25 subdigital lamellae on fourth digit of hindfoot.

C. Tokay *(Gekko gecko)* PAGE 86
Hawaiian Islands: possibly Oʻahu. Adults 120–150 mm SVL. Dorsum
brightly spotted with orange; tail nearly cylindrical without whorls of
pointed scales; digital pad with 17–24 subdigital lamellae on fourth digit
of hindfoot.

D. Micronesia Saw-tailed Gecko *(Perochirus ateles)* PAGE 109
Western Micronesia, southern Marianas, and Marshall Islands into
western half of the Federated States of Micronesia. Adults 55–88 mm
SVL. Stout flattened tail, strongly segmented, with ventrolateral fringe of
elongate projecting scales; digital pad with 16–21 subdigital lamellae on
fourth digit of hindfoot, distal 4 or 5 lamellae divided.

PLATE 7 | Geckos: House Geckos and Relatives

A. Common House Gecko *(Hemidactylus frenatus)*　　　PAGE 90
Pacific, widespread. Adults 42–60 mm SVL. Trunk with scattered
round tubercles among granular scales; elongate digital pads with
enlarged, medially divided subdigital lamellae, 8–11 lamellae on fourth
digit of hindfoot.

B. Fox Gecko *(Hemidactylus garnotii)*　　　PAGE 92
Pacific, Hawaiian Islands and French Polynesia westward. Adults 49–
66 mm SVL. Trunk with uniform granular scales, no tubercles present;
elongate digital pads with enlarged subdigital lamellae, only terminal 3
or 4 divided, 10–15 lamellae on fourth digit of hindfoot.

C. Palau Slender Gecko *(Hemiphyllodactylus ganoklonis)*   PAGE 93
Palau, most islands. Adults 28–34 mm SVL. Postsacral bar pinkish
yellow to bright orange; digital pads round with U-shaped subdigital
lamellae, 4 on fourth toe.

D. Indo-Pacific Slender Gecko *(Hemiphyllodactylus typus)*   PAGE 94
Pacific, Hawaii and French Polynesia westward. Adults 29–46 mm
SVL. Postsacral bar white to beige; digital pads round with U-shaped
subdigital lamellae, 5 on fourth toe.

PLATE 8 | Geckos: Round-tails

A. 'Eua Forest Gecko *(Lepidodactylus euaensis)* PAGE 95
Tonga, 'Eua. Adults 42–50 mm SVL. Usually dark brown face with
ruby brown lips; venter yellow suffused with brown.

B. Rotuma Forest Gecko *(Lepidodactylus gardineri)* PAGE 97
Rotuma. Adults 43–50 mm SVL. Face green to gray with gray lips; eye
encircled with yellow scales; venter yellow from chin onto tail.

C. Fiji Forest Gecko *(Lepidodactylus manni)* PAGE 100
Fiji, Kadavu, Ovalau, Viti Levu. Adults 35–48 mm SVL. Trunk with
five dark brown chevrons dorsally; venter uniform yellow from chin
onto tail.

D. Vanuatu Ant-nest Gecko *(Lepidodactylus buleli)*　　PAGE 95
Vanuatu, Espiritu Santo. Adults circa 38 mm SVL. Lips and postorbital
stripe golden tan; venter white with yellowish chin.

E. Solomon Forest Gecko *(Lepidodactylus guppyi)*　　PAGE 98
Solomon Islands, most islands. Adults 36–54 mm SVL. Face usually
without markings; venter yellowish white with yellow chin.

F. Vanuatu Forest Gecko *(Lepidodactylus vanuatuensis)*　　PAGE 105
Vanuatu, likely most islands. Adults 33–47 mm SVL. Trunk with five to
six dark brown transverse bars; venter creamy white from chin onto tail.

PLATE 9 | Geckos: Flat-tails

A. Mourning Gecko clone A *(Lepidodactylus lugubris)*   PAGE 99
Pacific, widespread. Adults 33–49 mm SVL. Dorsally trunk with tan to
beige background and series of narrow darker brown chevrons.

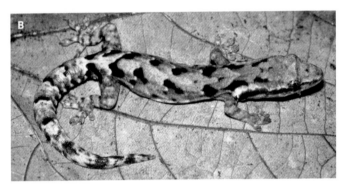

B. Mourning Gecko clone B *(Lepidodactylus lugubris)*   PAGE 99
Adults 33–45 mm SVL. Dorsally trunk with gray background and
series of large dark brown dorsolateral blotches.

C. Mourning Gecko clone C *(Lepidodactylus lugubris)*   PAGE 99
Adults 33–43 mm SVL. Dorsally trunk with reddish brown background and series of lighter brown narrow chevrons.

D. Mourning Gecko clone E *(Lepidodactylus lugubris)* PAGE 99
Adults 34–41 mm SVL. Dorsally trunk light beige to tan background,
usually unmarked.

E. Micronesia Flat-tailed Gecko *(Lepidodactylus moestus)* PAGE 101
Micronesia from Palau to Arno Atoll. Adults 32–41 mm SVL. Dorsally
trunk with light to medium brown background covered with numerous
dark brown spots; venter unicolor white to beige.

PLATE 10 | Geckos: Ground Dwellers and Hawaiian Aliens

A. Melanesia Slender-toed Gecko *(Nactus multicarinatus)* PAGE 107
Vanuatu and Solomon Islands. Adults 43–63 mm SVL. Chin with large postmental scale on each side of mental scale.

B. Pacific Slender-toed Gecko *(Nactus pelagicus)* PAGE 108
Western Pacific, most islands. Adults 48–75 mm SVL. Chin with no or very small postmental scale on each side of mental scale.

C. Orange-spotted Daygecko *(Phelsuma guimbeaui)* PAGE 114
Hawaiian Islands: Oʻahu. Adults 57–70 mm SVL. Dorsal ground color green and dorsolateral red stripe from above eye onto anterior trunk.

D. Golddust Daygecko *(Phelsuma laticauda)*   PAGE 115
Hawaiian and Society Islands. Adults 43–60 mm SVL. Dorsal ground color brown to green with heavy sprinkling of tiny golden yellow spots on side of neck and anterior trunk and usually three red streaks dorsally on posterior trunk.

E. Madagascar Giant Daygecko *(Phelsuma grandis)*   PAGE 112
Hawaiian Islands: Maui, Moloka'i and O'ahu. Adults 100–200 mm SVL. Dorsal ground color green with cape of red spots from neck onto anterior trunk.

PLATE 11 | Geckos: East Pacific

A. Shield-headed Gecko *(Gonatodes caudiscutatus)* PAGE 124
Galápagos, Baltra, San Cristóbal. Adults 36–42 mm SVL. Trunk with light-colored middorsal stripe that forks at the neck; commonly a white, centered spot on side behind shoulder.

B. Santa Fé Leaf-toed Gecko
(Phyllodactylus barringtonensis)
PAGE 116
Galápagos, Santa Fé. Adults ~40–42 mm SVL. Trunk without middorsal stripe; hindlimbs with no tubercles on thigh or crus (lower leg).

C. Pinta Leaf-toed Gecko
(Phyllodactylus baurii) PAGE 117
Galápagos, Española, Floreana, Pinta. Adults 33–42 mm SVL. Trunk with cream middorsal stripe; hindlimbs with no tubercles on thigh or crus.

D. Galápagos Leaf-toed Gecko
(Phyllodactylus galapagensis)
PAGE 119
Galápagos, widespread. Adults
33–46 mm SVL. Trunk with no
middorsal stripe; hindlimbs with
no tubercles on thigh or crus.

D

E. Guayaquil Leaf-toed Gecko
(Phyllodactylus reissi) PAGE 122
Native of coastal Ecuador and
Galápagos, Santa Cruz. Adults
40–75 mm SVL. Trunk with tan to
gray middorsal stripe; hindlimbs
with tubercles on crus.

E

F. Malpelo Leaf-toed Gecko *(Phyllodactylus transversalis)* PAGE 123
Isla Malpelo. Adults 56–58 mm SVL. Trunk with broad, transverse brown bands;
hindlimbs with tubercles on crus.

PLATE 12 | Iguanas: Pacific and Aliens

A. Fiji Banded Iguana
(Brachylophus bulabula) PAGE 135
Fiji, western islands. Adults 136–193 mm SVL. Small to modest-sized dorsal crest spines; circular nasal scale yellow with circular nostril.

B. Lau Banded Iguana
(Brachylophus fasciatus) PAGE 136
Fiji, Lau Islands, and Tonga, Tongatabu. Adults ~140–182 mm SVL. Small to modest-sized dorsal crest spines; elliptical nasal scale tannish orange to cream with a flattened elliptical nostril.

C. Fiji Crested Iguana
(Brachylophus vitiensis) PAGE 137
Fiji, western fringe of Vanua Levu
and Viti Levu, adjacent islets.
Adults 180–250 mm SVL. Modest
to large dorsal crest spines; pentagonal nasal scale bright orangish
yellow overlaps onto adjacent
scales; broad elliptical nostril.

D. Green Iguana
(Iguana iguana) PAGE 141
Hawaiian Islands: Maui, Oʻahu
and Fiji: Koro, Qamea, Taveuni.
Adults 220–380 mm SVL. Large
dorsal crest spines from nape onto
tail; dorsal ground color highly
variable, usually grayish; large
circular scale on cheek below
tympanum.

PLATE 13 | Iguanas: Galápagos

A. Marine Iguana
(Amblyrhynchus cristatus)

PAGE 134

Galápagos, widespread. Adults 220–548 mm SVL. Modest to large dorsal crest spine on nape and anterior trunk, smaller posteriorly; dorsal ground color mostly black.

B. Marine Iguana
(Amblyrhynchus cristatus)

PAGE 134

C. Santa Fé Land Iguana *(Conolophus pallidus)*   PAGE 139
Galápagos, Santa Fé. Adults 325–375 mm SVL. Low dorsal crest spines on nape and anterior trunk; dorsal ground color beige to light brown.

D. Galápagos Land Iguana
(Conolophus subcristatus)
PAGE 140
Galápagos, widespread. Adults
360–480 mm SVL. Low dorsal
crest spines on nape and anterior
trunk; dorsal ground color golden
brown.

E. Pink Land Iguana
(Conolophus marthae) PAGE 138
Galápagos, Volcan Wolf on Isla
Isabela. Adults 350–470 mm SVL.
Small dorsal crest spines; dorsal
ground color predominantly pink.

PLATE 14 | Lava and Tree Lizards: East Pacific

A. Galápagos Lava Lizard
(*Microlophus albemarlensis*) male   PAGE 145

Galápagos, widespread. Adults 75–95 mm SVL. Dorsal ground color medium brown with three broad cream-colored stripes in juveniles and subadult females.

B. Galápagos Lava Lizard
(*Microlophus albemarlensis*) female   PAGE 145

C. San Cristóbal Lava Lizard   **(*Microlophus bivittatus*)**   PAGE 146

Galápagos, San Cristóbal. Adults 65–75 mm SVL. Dorsal ground color medium brown with four broad cream-colored stripes.

D. Española Lava Lizard
(Microlophus delanonis)   PAGE 147
Galápagos, Española. Adults 84–138 mm SVL. Dorsal ground color
olive brown with six to eight longitudinal rows of dark brown spots.

E. Floreana Lava Lizard   *(Microlophus grayii)*   PAGE 149
Galápagos, Floreana. Adults 60–70 mm SVL. Dorsal ground color
medium brown with dark brown spots and crossbars.

F. Socorro Treelizard   *(Urosaurus auriculatus)*   PAGE 143
Islas Revillagigedo, Socorro. Adults ~46–55 mm SVL. Dorsal ground
color bright blue; dorsum with alternating longitudinal bands of granular and larger keeled scales.

PLATE 15 | Anoles: East Pacific and Aliens

A. Malpelo Anole *(Anolis agassizii)* male   PAGE 127
Isla Malpelo. Adults 71–114 mm SVL. Dorsal ground color dark
brown to blue with numerous small light spots; small dewlaps in both
sexes.

B. Malpelo Anole
(Anolis agassizii) female
PAGE 127

C. Green Anole
(Anolis carolinensis)
PAGE 129
Hawaiian Islands: wide-
spread, and Guam. Adults
45–72 mm SVL. Dorsal
ground color bright green
to gray; medium dewlap,
largely pink.

D. Knight Anole *(Anolis equestris)* PAGE 130
Hawaiian Islands: O'ahu. Adults 136–179 mm SVL. Dorsal ground color dark green to brown with broad white stripe diagonally on anterior trunk; medium dewlap, largely pink.

E. Brown Anole *(Anolis sagrei)* PAGE 131
Hawaiian Islands: widespread. Adults 34–70 mm SVL. Dorsal ground color medium brown to grayish brown; large dewlap, bright orangish red.

F. Brown Anole
(Anolis sagrei) PAGE 131

PLATE 16 | Snake-eyed Skinks

A. Fiji Snake-eyed Skink *(Cryptoblepharus eximius)*   PAGE 155
Fiji, widespread. Adults 33–40 mm SVL. Silvery beige dorsolateral
stripe edged above in black or dark brown; broad dark brown to black
lateral stripe from face onto tail.

B. Fiji Snake-eyed Skink
(Cryptoblepharus eximius)
melanistic form   PAGE 155

**C. Vanuatu Snake-eyed
Skink**
*(Cryptoblepharus
novohebridicus)*   PAGE 157
Vanuatu, widespread.
Adults 29–37 mm SVL.
White dorsolateral stripe
above broad dark brown to
black lateral stripe; light to
bright blue tail.

D. Oceania Snake-eyed Skink
(*Cryptoblepharus poecilopleurus*)   PAGE 158

Pacific, Marianas southward to Rapa Nui. Adults 37–51 mm SVL.
Cream to tan dorsolateral stripe and no distinct dark lateral stripe.

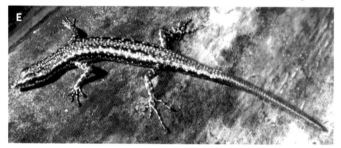

E. Palau Snake-eyed Skink   *(Cryptoblepharus rutilus)*   PAGE 159

Palau, most islands. Adults 33–38 mm SVL. White to cream dorsolateral and ventrolateral stripe edged above in black or dark brown;
broad dark brown to black lateral stripes enclose dark brown lateral
stripe on trunk.

PLATE 17 | *Emoia:* Widespread Striped Skinks

A. Pacific Blue-tailed Skink *(Emoia caeruleocauda)* adult   PAGE 165
Western Pacific, widespread. Adults 40–68 mm SVL. Digital lamellae broad; venter white; parietal eye absent; tail blue.

B. Pacific Blue-tailed Skink
(Emoia caeruleocauda) adult, stripes lost
PAGE 165

C

C. Pacific Blue-tailed Skink *(Emoia caeruleocauda)* juvenile PAGE 165

D. White-bellied Copper-striped Skink
(Emoia cyanura) PAGE 169
Central and western Pacific, widespread. Adults 39–56 mm SVL.
Digital lamellae narrow and numerous; venter white; parietal eye
present; tail green to brown (blue in northernmost populations).

E. Dark-bellied Copper-striped Skink
(Emoia impar) PAGE 171
Central and western Pacific, widespread. Adults 40–47 mm SVL.
Digital lamellae narrow and numerous; venter dusky to gray; parietal
eye absent; tail blue.

PLATE 18 | *Emoia:* Other Striped Skinks

**A. Papua Five-striped
Skink
*(Emoia jakati)*** PAGE 173
Palau to Marshall Islands.
Adults 37–53 mm SVL. Five
white stripes on dark brown
to black background.

**B. Red-lipped Striped
Skink
*(Emoia rufilabialis)*** female
PAGE 187
Solomon Islands, Santa
Cruz. Adults 46–65 mm
SVL. Three light stripes
on coppery to olive brown
background, stripes lost in
adult males; red lips.

**C. Red-lipped Striped
Skink
*(Emoia rufilabialis)*** male
PAGE 181

D. Taumako Skink
(Emoia taumakoensis)
female   PAGE 184
Solomon Islands, Taumako.
Adults 44–59 mm SVL.
Back with golden yellow
stripe on medium to dark
brown background, sides
lighter brown.

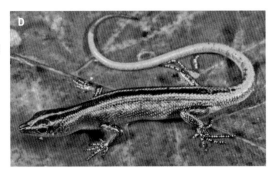

E. Taumako Skink
(Emoia taumakoensis) male
                    PAGE 184

PLATE 19 | *Emoia* and Other Skinks: Melanesian Robust Treeskinks

A. Anatom Treeskink *(Emoia aneityumensis)*   PAGE 161
Vanuatu, Anatom. Adults 71–96 mm SVL. Dorsal ground color light
to coppery brown, dorsally unicolor.

B. Erronan Treeskink *(Emoia erronan)*   PAGE 170
Vanuatu, Erronan. Adults 69–101 mm SVL. Head and body with
contrasting colors, either coppery red head and brown body or coppery
white head and black body.

C. Vanuatu Coppery Vineskink *(Emoia nigromarginata)*   PAGE 177
Vanuatu. Adults 52–76 mm SVL. Dorsal ground color coppery tan to
olive with grayish ivory dorsolateral stripes.

D. Toupeed Treeskink
(Emoia sanfordi)   PAGE 182
Vanuatu. Adults 92–114
mm SVL. Dorsal ground
color usually bright green;
various amounts of black
on top of head.

E. Emerald Treeskink
(Lamprolepis smaragdina)
PAGE 192
Solomon Islands, Palau, and
western Micronesia. Adults
80–107 mm SVL. Usually
overall bright green with
black edged scales, varying
to green anteriorly with
coppery brown posteriorly.

PLATE 20 | *Emoia:* Central Pacific Robust Treeskinks

A

A. Tonga Robust Treeskink
(*Emoia mokolahi*) PAGE 175
Tonga. Adults 86–105 mm
SVL. Light to medium
brown ground color, occa-
sional dark brown bars,
three dark spots on side
(neck, shoulder, midbody).

**B. Rotuma Barred
Treeskink
(*Emoia oriva*)** PAGE 178
Rotuma. Adults 73–90
mm SVL. Coppery brown
ground color with numer-
ous traverse series of near
black bars and numerous
lime-green tick marks.

**C. Pacific Robust
Treeskink
(*Emoia samoensis*)**
PAGE 182
Samoa. Adults 84–112
mm SVL. Coppery brown
ground color with some
black bars and few lime-
green tick marks.

D. Rarotonga Treeskink
(Emoia tuitarere) PAGE 187
Cook Islands. Adults 68–93
mm SVL. Coppery brown
ground color with numer-
ous traverse series of near
black bars and numerous
lime-green tick marks.

E. Fiji Barred Treeskink
(Emoia trossula) PAGE 186
Fiji. Adults 80–107 mm
SVL. Coppery brown
ground color with pattern
of few to moderate amounts
of dark brown bars and
yellow-green tick marks.

F. Mariana Skink
(Emoia slevini) PAGE 183
Mariana Islands. Adults
63–84 mm SVL. Medium
brown ground color usually
with dorsolateral row of
white spots from neck to
tail.

PLATE 21 | *Emoia* and *Prasinohaema:* Slender Treeskinks

A. Vitilevu Mountain Treeskink
(Emoia campbelli) PAGE 167
Fiji, Viti Levu. Adults 68–98 mm SVL. Medium brown ground color overlain with creamy yellow spots and small dark brown blotches.

B. Fiji Slender Treeskink
(Emoia concolor) PAGE 167
Fiji, widespread. Adults 58–95 mm SVL. Variable dorsal ground color, from uniform pale green to coppery tan or dull grayish green; small dark spotting infrequent and slight when present.

C. Vanualevu Slender Treeskink
(Emoia mokosariniveikau)
PAGE 176
Fiji, Vanua Levu. Adults circa 55 mm SVL. Coppery brown ground color with turquoise sides from jowl to hindlimb.

D. Green-bellied Vineskink
(Emoia cyanogaster)
PAGE 168
Solomon Islands, Vanuatu.
Adults 62–92 mm SVL.
Green lower sides and
venter.

**E. Polynesia Slender
Treeskink**
(Emoia tongana) PAGE 185
Samoa, Tonga, Wallis-
Futuna. Adults 53–75 mm
SVL. Coppery olive to tan
ground color, lightly speck-
led with brown; underside
light yellowish green.

**F. Green-blooded
Vineskink**
(Prasinohaema virens)
PAGE 200
Solomon Islands, wide-
spread. Adults 46–52 mm
SVL. Light to olive green
ground color with diffuse
yellow middorsal stripe.

PLATE 22 | *Emoia* **and Other Skinks: Unicolor Skinks**

A. Admiralty Brown Skink *(Carlia ailanpalai)*   PAGE 152

Southern Mariana Islands. Adults 46–59 mm SVL. Forefoot with four digits; unicolor medium brown or with darker brown sides.

B. Micronesia Black Skink *(Emoia arnoensis)*   PAGE 162

Central and eastern Melanesia, Chuuk to Arno Atoll and Nauru. Adults 73–86 mm SVL. Unicolor black; brown iris; medium interparietal scale.

C. Micronesia Spotted Skink *(Emoia boettgeri)* PAGE 164
Central and eastern Melanesia, Chuuk to Arno Atoll. Adults 60–77
mm SVL. Dorsally either unicolor olive brown or spotted with small
black spots.

D. South Pacific Black Skink *(Emoia nigra)* PAGE 176
Samoa, Tonga westward into Vanuatu. Adults 85–121 mm SVL.
Unicolor black; red iris; small interparietal scale.

PLATE 23 | *Emoia:* Two-toned Skinks

A. Striped Small-scaled Skink
(*Emoia adspersa*) PAGE 160
Central Pacific. Adults
63–85 mm SVL. Numerous
small scales on body;
coppery to orangish brown
ground color.

B. Seaside Skink
(*Emoia atrocostata*)
PAGE 163
Western Pacific, widespread.
Adults 57–98 mm SVL.
Two populational ground
colors, gray or brown; back
lighter than sides.

C. Seaside Skink
(*Emoia atrocostata*)
PAGE 163

D. Olive Small-scaled Skink
(Emoia lawesii)   PAGE 174
Samoa to Niue. Adults 77–106 mm SVL. Numerous small scales on body; medium to dark brown ground color, usually darker on sides.

E. Fiji Copper-headed Skink   *(Emoia parkeri)*   PAGE 179
Fiji, Kadavu, Ovalau, Taveuni, Viti Levu. Adults 43–52 mm SVL. Dorsal ground color usually coppery brown, strongly coppery on head; dark brown lateral stripe from head to hindlimb.

F. Pohnpei Skink   *(Emoia ponapea)*   PAGE 180
Pohnpei. Adults 43–51 mm SVL. Dorsal ground color medium brown with scattering of dark brown blotches forming an irregular dorsolateral stripe on each side.

PLATE 24 | Ground Skinks: Long Bodies and Short Limbs

**A. Speckled Litter Skink
(Caledoniscincus
atropunctatus)** PAGE 191
Vanuatu, southern islands.
Adults 38–53 mm SVL.
Dorsally medium brown
from head to tail, laterally
darker brown.

**B. Barred Recluse Skink
(Eugongylus
albofasciolatus)** PAGE 188
Solomon Islands, Palau, and
Micronesia from Chuuk to
Marshall Islands. Adults
103–150 mm SVL. Dorsal
ground color medium to
chocolate brown with
transverse bars of speckled
white.

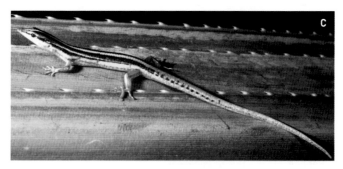

C. Pandanus Moth Skink *(Lipinia leptosoma)* PAGE 198
Palau, Babeldoad, Ngeaur, Oreor. Adults 35–44 mm SVL. Middorsal
stripe from snout to tail, white on head becoming yellow on neck and
body.

D. Pacific Moth Skink *(Lipinia noctua)*   PAGE 199
Central and western Pacific. Adults 37–47 mm SVL. Middorsal stripe beginning with bright white to cream parietal spot, becoming beige on body and base of tail.

E. Ono-i-Lau Ground Skink *(Leiolopisma alazon)*   PAGE 195
Fiji, Ono-i-Lau. Adults 43–60 mm SVL. Red snout grading into medium brown on body.

PLATE 25 | Ground Skinks: Long Bodies and Short Limbs

A. Micronesia Multi-keeled Sunskink
(*Eutropis multicarinata*)   PAGE 196
Palau, most islands. Adults 62–76 mm SVL. Ground color medium
brown, unicolor in adult males; light dorsolateral stripes in juveniles
and young females; white to light blue lip stripe.

B. Palau Ground Skink   (*Sphenomorphus scutatus*)   PAGE 201
Palau, most islands. Adults 31–44 mm SVL. Ground color light to
medium brown with small dark spots from neck onto tail.

C. Solomon Ground Skink *(Sphenomorphus solomonis)* PAGE 202
Northern New Guinea through Solomon Islands. Adults 50–57 mm
SVL. Dorsally light to medium brown with small dark spots on neck
and trunk.

D. Malpelo Galliwasp *(Diploglossus millepunctatus)* PAGE 204
Isla Malpelo. Adults 190–240 mm SVL. Dorsal ground color dark
brown to black, speckled with numerous tiny yellowish white spots.

PLATE 26 | Lizards: Monitor and Aliens

A. Pacific Monitor
(Varanus indicus) PAGE 205
Western Micronesia. Adults
275–580 mm SVL. Long
slender neck; dorsal ground
color black with light spots.

B. Pacific Monitor
(Varanus indicus) PAGE 205

C. Veiled Chameleon
(Chamaeleo calyptratus)
PAGE 209

Hawaiian Islands: Maui.
Adults 180–210 mm SVL.
Large sharkfin-like casque
on rear third of head.

D. Jackson's Chameleon
(Trioceros jacksonii)
PAGE 210

Hawaiian Islands: most
islands. Adults 70–140 mm
SVL. Three horns project-
ing horizontally from top
of head.

E. Garden Skink
(Lampropholis delicata)
PAGE 194

Hawaiian Islands: most
islands. Adults 31–46 mm
SVL. Dorsal ground color
medium to orangish brown
with dark brown mottling.

PLATE 27 | Blindsnakes

A. Palau Blindsnake
(Ramphotyphlops
acuticaudus)   PAGE 213
Palau, Babeldaob, and
Oreor. Adults 154–245 mm
SVL. Strongly bicolor, dark
brown above and yellow-
ish tan below; moderately
robust body.

B. Brahminy Blindsnake
(Ramphotyphlops braminus)
PAGE 215
Pacific, widespread. Adults
95–175 mm SVL. Shiny
black above and dark gray
below; very slender, nearly
thread-like body.

C. Melanesia Blindsnake
(Ramphotyphlops
depressus)   PAGE 216
Solomon Islands, wide-
spread. Adults 180–220 mm
SVL. Shiny yellowish brown
above and pale yellow
below; robust body.

D. Ant Atoll Blindsnake *(Ramphotyphlops adocetus)* PAGE 214
Micronesia, Pohnpei State, Ant Atoll. Adults 154–390 mm SVL.
Strongly bicolor, dark brown above and tan below; moderately robust
body.

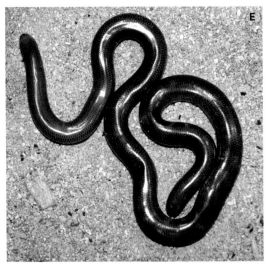

E. Ulithi Atoll Blindsnake
(Ramphotyphlops hatmaliyeb) PAGE 216
Micronesia, Yap State, Ulithi. Adults 178–416 mm SVL.
Strongly bicolor, reddish brown above and pigmentless
below; robust body.

PLATE 28 | Boas

A. Pacific Treeboa *(Candoia bibroni)* PAGE 218
Solomon Islands eastward to Samoa and Tonga. Adults 460–1,460
mm SVL. Body scales on trunk in longitudinal rows; keels parallel to
longitudinal axis; subocular scale separates eye from upper lip scales.

B. Pacific Treeboa *(Candoia bibroni)* PAGE 218

C. Melanesia Bevel-nosed Boa *(Candoia paulsoni)* PAGE 219
Northern New Guinea and Solomon Islands. Adults ~650–1,365 mm total length. No subocular scale, eye touches upper lip scales; underside of tail white with dark blotches.

D. Palau Bevel-nosed Boa *(Candoia superciliosa)* PAGE 220
Palau, widespread. Adults ~400–800 mm SVL. No subocular scale, eye touches upper lip scales; underside of tail with a dark bordered elongate and white spot immediately behind vent.

PLATE 29 | Colubrid Snakes: East Pacific

A. Galápagos Racer
(Alsophis biseralis)

PAGE 222

Galápagos, widespread.
Adults 483–1,005 mm SVL.
Variable dorsal coloration
from unicolor dark brown
to light brown stripes and
spots on dark background;
venter dusky brown.

B. Galápagos Racer
(Alsophis biseralis)

PAGE 222

**C. Galápagos Banded
Snake**
(Antillophis slevini)

PAGE 222

Galápagos, Fernandina,
Isabela, Pinzón. Adults
~345–420 mm SVL.
Dorsally dark brown bands
on yellowish brown back-
ground; ventrally lighter
with each ventral scale
narrowly dark edged.

D. Galápagos Striped Snake
(Antillophis steindachneri)
PAGE 223
Galápagos, Baltra, Rábidan, Santa Cruz, Santiago. Adults 335–405 mm SVL. Dorsally two broad, cream colored stripes on light to medium brown background; ventrally alternating bars of light and dark.

E. Española Racer
(Philodryas hoodensis)
PAGE 227
Galápagos, Española, Gardner. Adults 468–856 mm SVL. Narrow white dorsolateral stripes, from eye to posterior third of trunk, on dark brown background; ventral scales white, usually with a dark spot laterally on each side.

PLATE 30 | Land Snakes: West Pacific

A. Brown Treesnake
(Boiga irregularis)

PAGE 223

Guam and Saipan. Adults 800–2400 mm SVL. Slender bodied with large head distinct from neck.

B. Solomon Treesnake
(Dendrelaphis salomonis)

PAGE 225

Solomon Islands, widespread. Adults ~850–1,450 mm SVL. Middorsal row of enlarged scales; each ventral scale folded upward on lateral edge.

C. Palau Treesnake
(Dendrelaphis striolatus)

PAGE 225

Palau, widespread. Adults 608–982 mm SVL. Middorsal row of enlarged scales; each ventral scale folded upward on lateral edge.

D. Dog-faced Mudsnake
(Cerberus dunsoni)
PAGE 237
Palau, widespread. Adults
560–900 mm SVL.
Dorsally unicolor gray to
dusky brown with darker
brown bands; dorsal
scales strongly keeled
(unicarinate).

E. Fiji Bola
(Ogmodon vitianus) subadult
PAGE 235
Fiji, Viti Levu. Adults 205–
325 mm SVL. Dorsally
unicolor shiny black;
juveniles and subadults
with white parietal bar.

PLATE 31 | Seasnakes and Seakraits

A. Pacific Yellow-banded Seasnake
(Hydrophis coggeri)
PAGE 229
Vanuatu and Fiji. Adults ~900–1,265 mm SVL. Ventral scales width about half or less that of terrestrial snakes; dorsal ground color yellowish olive with dark olive bands; black tail tip.

B. Yellow-lipped Seakrait
(Laticauda colubrina)
PAGE 231
Western Pacific, including western Micronesia, Samoa, and Tonga. Adults 295–1,655 mm SVL. Ventral scales not greatly reduced in width from that of terrestrial snakes; dorsum bluish with all black bands circling body; yellow tail tip.

C. Vanuatu Yellow-lipped Seakrait
(Laticauda frontalis)
PAGE 232
Vanuatu, Efate, Espiritu Santo and likely islands between. Adults 293–783 mm SVL. Ventral scales not greatly reduced in width from that of terrestrial snakes; dorsum bluish with black bands, only posterior ones encircle trunk; yellow tail tip.

D. Dark-lipped Seakrait
(Laticauda laticaudata)
PAGE 233
Vanuatu and Fiji. Adults
740–1,130 mm SVL.
Ventral scales not greatly
reduced in width from
that of terrestrial snakes;
dorsum bluish gray with
few to many broad black
bands that encircle body;
white tail tip.

**E. Yellow-bellied
Seasnake**
(Pelamis platura) PAGE 236
Pacific, widespread and
pelagic. Adults ~400–600
mm SVL. Ventral scales
greatly reduced in width
from that of seakraits and
nearly indistinguishable
from lateral trunk scales;
brown to black dorsally,
bright yellow ventrally.

PLATE 32 | Seaturtles

A. Loggerhead Seaturtle
(*Caretta caretta*) PAGE 241
Pacific, widespread. Adults 700–1,460 mm SCL. Large head; two pairs of prefrontal scales on snout; five pleural scutes on each side of carapace.

B. Green Seaturtle
(*Chelonia mydas*) PAGE 242
Pacific, widespread. Adults 680–1,080 mm SCL. Small head; single pair of prefrontal scales on snout; four pleural scutes on each side of carapace.

C. Hawksbill Seaturtle
(*Eretmochelys imbricata*)
PAGE 244
Pacific, widespread. Adults 680–930 mm SCL. Small head; two pairs of prefrontal scales on snout; four pleural scutes on each side of carapace.

D. Olive Ridley Seaturtle
(Lepidochelys olivacea)
PAGE 245

Eastern Pacific. Adults 580–750 mm SCL. Moderate-sized head; two pairs of prefrontal scales on snout; five to seven pleural scutes on each side of carapace.

E. Leatherback Seaturtle
(Dermochelys coriacea)
PAGE 247

Pacific, widespread. Adults 1250–1,600 mm SCL. Large head; skin of head and shell without horny scutes.

PLATE 33 | Freshwater Turtles: Aliens

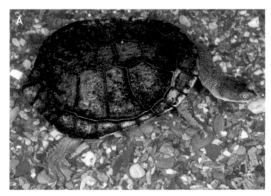

A. Eastern Long-necked Turtle
(Chelodina longicollis)
                              PAGE 239
Cook Islands, Rarotonga.
Adults ~160–240 mm
SCL. Extremely long
neck, retracted by laying it
sideways beneath anterior
carapace margin.

B. Snapping Turtle
(Chelydra serpentina)
                              PAGE 249
Guam. Adults 180–490 mm
SCL. Long tail; posterior
edge of carapace strongly
serrated.

C. Chinese Three-keeled Pondturtle
(Mauremys reevesii)
                              PAGE 252
Guam. Adults 120–236 mm
SCL. Carapace with three
longitudinal ridges, center
one highest; few yellow
stripes on head and neck.

D. Chinese Stripe-necked Pondturtle
(Mauremys sinensis)
PAGE 254
Guam. Adults 107–270 mm SCL. Carapace depressed medially; head and neck with numerous narrow yellow stripes.

E. Red-eared Slider
(Trachemys scripta)
PAGE 251
Hawaiian Islands, French Polynesia, Bonin Islands, southern Marianas, and Guam. Adults 100–250 mm SCL. Carapace rounded or domed medially; red postorbital stripe from eye to anterior neck.

PLATE 34 | Galápagos Tortoises

A. Santiago Giant Tortoise
(Chelonoidis darwini)
PAGE 258
Galápagos, Santiago. Adults
55–140 cm SCL. Shell
shape intermediate between
domed and saddle-back;
gray to black carapace.

B. Pinzón Giant Tortoise
(Chelonoidis ephippium)
PAGE 258
Galápagos, Pinzón. Adults
46–87 cm SCL. Saddle-back
shell shape; brownish gray
carapace.

C. Alcedo Giant Tortoise
(Chelonoidis vandenburghi)
PAGE 262
Galápagos, Volcan Aledo
of Isabela. Adults ~65–145
cm SCL. Shell shape domed;
gray to black carapace.

D. Sierra Negra Giant Tortoise
(Chelonoidis guentheri)
PAGE 259
Galápagos, Volcan Sierra
Negra of Isabela. Adults
60–120 cm SCL. Shell
shape intermediate between
domed and saddle-back;
gray brown to black
carapace.

E. La Caseta Giant Tortoise
(Chelonoidis porteri)
PAGE 261
Galápagos, Santa Cruz.
Adults 65–150 cm SCL.
Domed shell shape; black
carapace.

F. Cerro Azul Giant Tortoise
(Chelonoidis vicina)
PAGE 263
Galápagos, Volcan Cerro
Azul of Isabela. Adults
54–125 cm SCL. Shell
shape intermediate between
domed and saddle-back;
black carapace.

PLATE 35 | Softshell Turtles and Saltwater Crocodile

A. Spiny Softshell
(Apalone spinifera)

PAGE 264

Hawaiian Islands: Oʻahu. Adults 80–540 mm SCL. Carapace slightly thickened on anterior margin, entire surface lightly spiny; neck skin smooth.

B. Wattle-necked Softshell
(Palea steindachneri)

PAGE 265

Hawaiian Islands: Kauaʻi, Oʻahu. Adults 200–450 mm SCL. Carapace with thick ridge on anterior margin, entire surface smooth; cluster of long papillae on top and side of neck.

C. Chinese Softshell adult
(Pelodiscus sinensis)

PAGE 266

Hawaiian Islands: Kauaʻi, Maui, Oʻahu. Adults ~200–330 mm SCL. Carapace slightly thickened on anterior margin, entire surface smooth; neck skin smooth.

**D. Chinese Softshell
hatchling
(*Pelodiscus sinensis*)**
PAGE 266

**E. Saltwater Crocodile
(*Crocodylus porosus*)**
PAGE 267
Southwest Pacific islands.
Adults 38 cm – >3 m SVL.
Single cluster of five or six
osteoderms on neck widely
separated from dorsal and
cranial armor.

FROGS

Frog is a frog is a frog is a frog, to borrow from Gertrude Stein's line "Rose is a rose is a rose is a rose" in her poem "Sacred Emily." Having seen one frog, we are able to recognize all other frogs as frogs. Frogs are unique among vertebrates in having a short tailless body, the head seemingly attached directly to the body without a neck, and the fore-limbs smaller than the hindlimbs. Most frogs move by bipedal jump-ing, powered by strong, robust hindlimbs. This form of jumping loco-motion is assumed to have driven the evolution of the unique froggy (anuran) body form. However, not all frogs are jumpers and even the jumpers can walk.

Frogs (Anura) comprise one of three major groups of living amphib-ians and are the only amphibians to have successfully colonized a few oceanic islands in the Pacific. All amphibians breathe mainly through their skin (cutaneous respiration), and oxygen transfer from air to blood requires an aqueous surface. Oxygen must first diffuse into water then diffuse across the skin interface to the blood; thus the skin is highly permeable, in both directions, to gas and water. This permeability places frogs at risk of rapid water loss in dry air or in saltwater, hence they are generally unable to cross saltwater barriers. The presence of native frogs in Fiji and Palau remains unexplained; all other frogs on Pacific islands arrived with the assistance of humans, all within the last century.

Bufonidae: Toads

Bufonidae is a diverse family comprising more than 40 genera and over 500 species. Toads occur naturally worldwide except in oceanic islands and the Australopapuan region. Most toads have short, stout bodies and short, strong limbs. They typically walk or hop and are predomi-nantly terrestrial creatures. Many species have prominent glands on the neck and rear of the head, and they often have "warts" over much of the top of the body and limbs. These structures are concentrations of poison glands that are numerous even in the smoother portions of the skin. The glandular secretions are highly toxic and protect toads from most vertebrate predators. As a group, toads have a broad physiologi-cal tolerance, living in areas with conditions that range from subzero winters to brutally hot and dry summers.

tion of this genus encompasses the entire range of the family, occurring in the Philippines and New Guinea eastward into the Pacific (Palau, the Solomon Islands, and Fiji).

Platymantis pelewensis   **Palau Frog**

Plate 1

APPEARANCE: Midsize frog with dorsal skin lightly tuberculate, bearing modest number of narrow longitudinal glandular ridges (folds) of variable length from rear of head to sacrum. Ridges disappear laterally, and skin becomes more tuberculate with small pebble-like surface that continues onto venter. Coloration varies from somber, nearly uniform medium-brown to brightly colored with different shades of brown. Bolder patterns often consist of broad, light-brown middorsal area and pair of dorsolateral stripes. Fore- and hindlimbs are often alternately barred in light and dark brown. Face typically bears dark-brown mask, lips alternately barred beneath eye, and dark-brown parenthesis-shaped mark extends from eye to behind tympanum. Venter is usually light brownish white. Fore- and hindlimbs are moderately robust and of medium length; forefeet are web-free and hindfeet lightly webbed at base. Digit tips are barely expanded on forefeet, they are expanded slightly more on hindfeet; nonetheless, each digit is tipped with small circular pad.

SIZE: Adult females average larger than males. Adult females range from 36 to 54 mm SVL (mean 46 mm); males from 34 to 40 mm SVL (38 mm).

OCCURRENCE: This endemic frog occurs widely throughout Palau from Kayangel to Ngeaur.

HABITAT CHOICE: Moisture, as for most frogs, is essential. During moist weather, its habitat ranges from the forest into towns and villages and other highly disturbed areas such as grasslands and scrub if moist hideaways are present that allow it to survive dry episodes. In the 1990s, Crombie and Pregill found large congregations of adult females in natural caves and Japanese bunkers. They noted that these retreats retain moisture throughout the year and have an abundant food source of cave crickets.

REPRODUCTION: External fertilization (amplexus) and direct development. Females lay clusters of about 30 eggs. Development is fast, and tiny froglets hatch in about three weeks. Only a few clutches of eggs have been discovered; unfortunately, the discoverer did not report where he found them or whether a parent was in attendance (likely). Males call year-round, although in greater numbers during wet periods. They call from beneath detritus, usually from small depressions or calling chambers. The call is likened to a high-pitched chatter of five to ten notes.

Platymantis vitiana   **Fiji Ground Frog**

Plate 1

APPEARANCE: Midsize to large frog with smooth, shiny skin above and below. Coloration is somber nearly uniform medium brown, occasionally reddish or tannish brown. Most individuals have light-colored spot behind tympanum. Fore- and hindlimbs are uniform brown, matching body coloration. Venter is dusky with fine mottling of medium brown on white to light tan background. Fore- and hindlimbs are moderately robust and of medium length; fore- and hindfeet are web-free. Digit tips are barely expanded on both fore- and hindfeet; nonetheless, each digit is tipped with small circular pad barely wider in its middle than its base.

SIZE: Adult females average larger than males. Adult females range from 60 to 110 mm SVL; males from 40 to 60 mm SVL.

OCCURRENCE: *P. vitiana* is a Fijian endemic that once occurred widely among the northern main islands of Viti Levu and Vanua Levu and associated lesser islands. Introduction of the mongoose in the mid-19th century extirpated the Ground Frog from Viti Levu and Beqa. Populations survive at low densities on Ovalau, Gau, Taveuni, Vanua Levu, and Viwa.

HABITAT CHOICE: *P. vitiana* is a forest-floor inhabitant and prefers canopied forest with good forest-floor litter. They also occur in gardens, such as kava patches, streamside in hill streams with only modest gallery forest, and even beachside.

REPRODUCTION: External fertilization (amplexus) and direct development. Females lay clusters of about 50 eggs in nests beneath or within rotting logs. Apparently, neither females nor males remain with the eggs. Eggs develop quickly and hatch in approximately 30 days. Metamorphosed froglets remain beneath logs, although the actual length of this residence is unknown. Because there is no evidence of males using stationary calling sites, females and males probably locate one another by scent or sight.

MISCELLANEA: Both sexes produce alarm calls. The female's call is a dog-like bark; males have a softer chirping sound. Perhaps this latter call also serves to identify males and attract females. The Ground

Frog is most active during evenings of light to modest rains versus dry periods or heavy rains.

Another Ground Frog has recently been found on Taveuni. Details of its morphology other than its size are limited. This frog is about 1.5 times larger than the sympatric *P. vitiana*.

Platymantis vitiensis   **Fiji Treefrog**

Plate 1.

APPEARANCE: Midsize frog with smooth, shiny skin above and below. Coloration is incredibly variable with over two dozen morphotypes regularly observed. Ground colors range from shades of brown, gray, and yellow to orange. Patterns of different colors contrast with a darker background and include differently shaped large blotches on the back or a narrow to broad light-colored middorsal stripe. Fore- and hindlimbs have contrasting light- and dark-brown barring to barely visible barring. Face typically has dark-brown canthal stripe and, behind eye, dark-brown parenthesis-shaped mark over and curving behind tympanum. Venter is usually uniformly and lightly colored, and can match or differ from dorsal ground color. Fore- and hindlimbs are moderately robust and of medium length; fore- and hindfeet are web-free. Digit tips are strongly expanded on forefeet, less so on hindfeet. Digital pads of forefeet are triangular with rounded edges; hindfeet pads are similarly shaped but distinctly smaller.

SIZE: Adult females average larger than males. Adult females range from 47 to 60 mm SVL; males from 32 to 45 mm SVL.

OCCURRENCE: *P. vitiensis* is a Fijian endemic. It occurs on the main islands of Viti Levu, Vanua Levu, Ovalau, and Taveuni.

HABITAT CHOICE: This treefrog lives in a variety of forests. It is often more abundant in streamside trees and shrubs, but occurs widely through the forest as well. It also survives and reproduces in anthropogenic habitats, such as isolated palms and pandanus in pastures and trees and shrubs around rural gardens if enough moisture is available year-round.

REPRODUCTION: External fertilization (amplexus) and direct development. Females lay clusters of 20 to 40 eggs in the axils of pandanus, palms, and similar vegetation pockets that retain high moisture. Development is fairly rapid, with froglets hatching in about 30 days. Calling males produce a gentle call reminiscent of a dripping water faucet: two drips then a pause, repeated over and over again.

Dendrobatidae: Poison Frogs

Dendrobatids are a group of small to moderate-sized terrestrial frogs native to the Neotropics. Of the more than 280 species of Dentrobatidae, many warn predators of their highly toxic skin secretions with their bold colors. Their various common names—Poison-arrow Frogs, Dart-poison Frogs, or Poison Frogs—result from these skin secretions. All species—even the cryptic colored black, brown, and white species—have poisonous skin secretions, but the secretions that are extremely toxic to vertebrates are limited to a few species in the genus *Phyllobates*. All Poison Frogs are diurnal and lay their eggs on land. Some species display parental care, with one parent remaining with the developing eggs and guarding them, and then transporting the hatching larvae to water to complete their development. The guarding parent can be the female or the male; the sex of the resident parent is species-specific.

Dendrobates auratus   Green and Black Poison-dart Frog

Plate 4

APPEARANCE: Small frog with smooth glossy skin. Dorsal ground color is black with irregularly shaped markings of bluish green. Dorsal coloration ranges from near unicolor to variously shaped green blotches, mostly moderate sized. Underside is black and unicolor to green blotched. Fore- and hindlimbs are slender; hindlimbs are moderately long, forelimbs about two-thirds length of hindlimbs; forefeet are web-free and hindfeet nearly so with slight basal web between third and fourth and between fourth and fifth toes; all digits have moderately large terminal disks with distinctly truncate distal end.

SIZE: Hawaiian adult females average slightly larger than males. Adult females typically range from 21 to 32 mm SVL (mean 29 mm); males from 20 to 27 mm SVL (25 mm). In Costa Rica, adult females and males are nearly equal-sized and range from 25 to 42 mm SVL.

OCCURRENCE: *D. auratus* is an inhabitant of the humid Pacific and Caribbean lowlands of middle Central America to northern South America. They were introduced into the upper Manoa Valley of Oʻahu to assist in the control of mosquitoes in 1932. A small population persists in this valley. It is also occurs in residential gardens and alien wet forest on the windward side of the Koʻolau Range (Oʻahu) and Maui.

HABITAT CHOICE: This frog lives on the forest floor and is widespread in the forest during wet periods. Individuals concentrate around streams and seepages in the dry season.

REPRODUCTION: External fertilization and indirect development. Males attract females by a low, slurred, buzzing call to a nesting site, usually in a hole beneath forest-floor detritus or at the base of a tree. Courtship is unusual for frogs because the male does not amplex the female. Instead he courts her by touching or resting his forefeet on her head. She responds by laying a small clutch of eggs (5–13) and departing. The male fertilizes the eggs and remains with them. Once they hatch, in about a week, he allows the tadpoles to squirm onto his back; he then carries them to a small water container such as a tree hole or a pool in a fallen palm frond, where the larvae complete their development in an average of nine to ten weeks.

Dicroglossidae: Fork-tongued Frogs

This family of frogs was formerly classified as a subgroup of Ranidae. Most Fork-tongued Frogs have the "true-frog" or ranid body form, and the majority are aquatic or semi-aquatic. There are about a dozen genera of Fork-tongued Frogs and more than 200 species, ranging from small species of 20–22 mm SVL to large ones exceeding 100 mm SVL. All dicroglossids lay aquatic eggs that hatch into tadpoles—that is, all exhibit indirect development. The duration of the tadpole stage can be as short as several weeks or as long as several months. With the exception of a single African species, all these frogs are Asian, occurring from Pakistan to Japan and south into the Lesser Sunda Islands and the Philippines.

Fejervarya cancrivora   **Crab-eating Frog**

Plate 3, Figure 2

APPEARANCE: Midsize to large frog, with mostly smooth skin and few short longitudinal glandular ridges on back and smooth, shiny skin below. Dorsal ground color is medium brown with scattering of midsize and irregularly shaped dark-brown spots. Anterodorsal surface of thigh and dorsal surface of crus (lower leg) have dark-brown bars on medium-brown background. Venter from throat to vent and limbs are usually glossy white. Chin is dusky, and anterolateral surface of neck bears bilateral dark smudges (vocal sacs) in males. Fore- and hindlimbs are robust, forelimbs short and hindlimbs moderately long; forefeet are web-free or slight webbed at base of finger and hindfeet strongly webbed with web reaching distal end of penultimate phalange of fourth toe. All digits have bluntly rounded distal tips.

SIZE: Adult females average larger than males. In Guam, adult females range from 65 to 103 mm SVL; males from 60 to 76 mm SVL.

OCCURRENCE: Crab-eating Frogs apparently arrived in Guam in the early 2000s. Their presence was noted first in west-central Guam, and now this species occupies the southern two-thirds of the island. Elsewhere, it occurs from eastern coastal India eastward through southern Thailand to southeastern China, and southward through the Greater and Lesser Sundas to Flores. Its presence in the Philippines and few areas of northern New Guinea may be the result of human introduction.

HABITAT CHOICE: *F. cancrivora* seldom occurs away from slow-moving or stationary water, typically in open-canopied habitats such as marshes. Adults and tadpoles tolerate and live in brackish waters of coastal marshes and tidal streams. Although capable of living in brackish areas, it occurs more widely, including low montane habitats to ~1,000 meters.

REPRODUCTION: External fertilization (amplexus) and indirect development. Females lay clusters of 500 to 1,000 eggs in ponds and paddies or in areas of streams with low or no water movement. Eggs typically hatch in less than a week, and tadpole development lasts about two months. Males produce a rapid, throaty gargle mating call.

MISCELLANEA: Regional differentiation has not been examined in Crab-eating Frogs, although it likely has occurred. One indication of this is the size ranges given for different populations. For example, Javan *F. cancrivora* are reported to attain body lengths to 120 mm, although 100 mm SVL is the average.

Fejervarya "limnocharis"   **Paddy Frog**

Plate 3, Figure 2

APPEARANCE: Midsize frog with dorsal skin bearing numerous short to long longitudinal glandular ridges and few scattered glandular tubercles; skin smooth below. Dorsal ground color is medium to moderately dark brown with scattering of midsize and irregularly shaped dark brown spots; often narrow to broad middorsal light-brown stripe extends from tip of snout to vent. Anterodorsal surface of thigh and dorsal surface of crus bear dark-brown bars on medium-brown background. Venter from throat to vent and underside of limbs are usu-

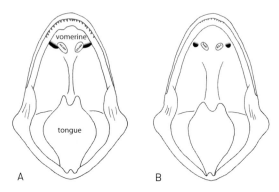

FIGURE 2. Mouth morphology of Guam's *Fejervarya*.
(A) *F. cancrivora*. (B) *F. "limoncharis."* The vomerine
ridge and teeth pattern differ between these two
sympatric species. In *F. cancrivora*, the bony ridge
supporting the vomerine teeth is longer and more
robust than in *F. "limoncharis."* Additionally, the teeth
are larger and more numerous in *F. cancrivora*.

ally glossy white; chin is dusky and anterolateral surface of neck bears
bilateral dark smudges (vocal sacs) in males. Fore- and hindlimbs are
robust, forelimbs short and hindlimbs moderately long; forefeet are
web-free or have slight webbing at the base of fingers, and hindfeet are
strongly webbed with web reaching distal end of penultimate phalange
of fourth toe. All digits have bluntly rounded distal tips.

SIZE: Adult females average larger than males. Adult females typically
range from 40 to 76 mm SVL (mean ~54 mm); males from 38 to 60 mm
SVL (~52 mm).

OCCURRENCE: Members of the *F. limnocharis* group occur widely in
southern Asia from peninsular India to southern China and Taiwan,
and southward in the Greater and Lesser Sundas.

Paddy Frogs were first reported from Guam in 2003 and now are
widespread throughout the southern half of the island. The origin of
these frogs is uncertain, and the assignment of a specific name is not
possible.

HABITAT CHOICE: *F. "limnocharis"* is a lowland species, mainly associated with open-canopied habitats such as marshes, canals, and flooded
agricultural lands. Unlike *F. cancrivora*, Paddy Frogs regularly move
away from standing-water habitats in rainy or high humidity weather.

REPRODUCTION: External fertilization (amplexus) and indirect development. The eggs are deposited as a surface film of about a thousand eggs

in standing water or areas of low or no water movement, commonly in marshes, paddies, and flooded fields. Eggs typically hatch quickly, and the tadpoles grow and metamorphose in six to eight weeks. Males call from the water's edge, commonly beneath overhanging vegetation. The call ranges from a rapid, raspy chirping of three to five seconds to a single, short, throatier "wah."

MISCELLANEA: In the 1960s, *F. "limnocharis"* was considered a single species ranging across southern Asia from Pakistan to Taiwan and Japan in the north and the Philippines in the south. At the beginning of 2010, this supposed pan-Asian species was well recognized as an extremely diverse species group, with more than 20 species now recognized and dozens more expected. True *Fejervarya limnocharis* occupy the southern half of the Malaya Peninsula and the Greater Sundas. The origin of the Guam population is unknown and will require a thorough molecular analysis of *"limnocharis"* populations of eastern Asia to determine its origin or multiple origins.

Eleutherodactylidae: Rain Frogs

Eleutherodactylid frogs are a speciose group of four genera of Neotropical frogs with most species living in forest habitats. They are terrestrial to arboreal species, and of the 200 plus species, most are West Indian and members of the genus *Eleutherodactylus*. Most species have restricted distributions on their native island and often small distributions even on modest-sized islands. Only a few species—*E. coqui, E. johnstonei,* and *E. planirostris*—have proved to be successful invaders. All eleutherodactylids display direct development, laying eggs in moist terrestrial sites or in bromeliads or leaf axils in arboreal sites. The developing embryo stays within the egg capsule and hatches in about three weeks as a miniature froglet, totally skipping the tadpole stage. Rain Frogs range from tiny adults of 20–22 mm SVL to moderate-sized ones of 60–70 mm SVL.

Eleutherodactylus coqui   Coquí

Plate 4

APPEARANCE: Midsize frog with finely tuberculate or rugose skin. Dorsal ground color ranges from tan or light brown to moderately dark brown, with variety of patterns from near unicolor (most frequent pattern in the Hawaiian Islands) and variously striped to moderately mottled or small spotted. Various striped patterns include middorsal stripe, usually narrow, from snout to vent, and moderately broad, light-colored dorsolateral stripe on each side of the trunk; middorsal and dorsolateral stripes can occur together or singularly on an individual. Most abundant patterns other than stripes consist of marks on dorsum; these marks can be dark chevron, W-shaped, or pair of curved marks over shoulders. A dark horizontal stripe is usually present from snout through eye extending above tympanum and curving downward to forelimb insertion. Underside is commonly pale gray to whitish, occasionally yellow to orange, with varying density of fine, dark stippling. Fore- and hindlimbs are slender and long; fore- and hindfeet are web-free. All digits have large terminal disks whose distal end is more square than rounded.

SIZE: Adult females average larger than males. Adult females attain a maximum length of ~63 mm SVL, although their body size typically ranges from 39 to 58 mm SVL (mean ~46 mm); males from 29 to 43 mm SVL (~36 mm).

OCCURRENCE: Until recently, the Coquí occurred widely in the Hawaiian Islands. It has been eradicated from Oʻahu and Maui; eradication treatment on Kauaʻi is ongoing; and only in Hawaiʻi do populations persist. They arrived in the islands on nursery stock, initially establishing pocket populations around plant nurseries and then spreading outward. A Coquí was heard around a Guam nursery in 2003, but fortunately the species did not become established there. *E. coqui* is a native of Puerto Rico, where locals gave it the vernacular name from the male's mating call. The presence of the Coquí on other West Indian islands probably derives from introductions.

HABITAT CHOICE: *E. coqui* is an ecological generalist capable of surviving and reproducing in a variety of habitats, from overgrown urban lots to primary forest, in sites that retain pockets of moisture and are never totally dry. Terrestrial during the day and staying beneath rocks and ground litter, they emerge at night and climb into shrubs and even high into trees to forage.

REPRODUCTION: Coquís have internal fertilization and direct development. Females lay 20 or more eggs (mean ~26) in a site selected by the male. Clutch size appears geographically variable and has not been determined for Hawaiian populations. The male attracts a female to his nest site, usually a rolled leaf or detritus crevices, for egg deposition. Courtship involves a nontypical amplexus where the male's and female's vents touch and the male deposits sperm directly into the female's cloaca.

She deposits the fertilized eggs in to the nest and departs; he remains to guard the eggs from predators and desiccation until they hatch. Froglets hatch in 17 to 26 days. In Puerto Rico, reproduction occurs year-round and is greatest during periods of high humidity and temperature.

MISCELLANEA: These frogs consume a variety of small invertebrates. Alien populations can become exceedingly dense, with more than 20,000 individuals per hectare. At these densities, the male's loud, two-note call, "co-quí," constantly repeated in large choruses, becomes a cacophony.

Eleutherodactylus planirostris   Greenhouse Frog

Plate 4

APPEARANCE: Small frog with weakly tuberculate skin. Dorsal ground color ranges from green to moderately dark brown. Dorsal coloration displays two patterns, mottled and striped (rare in Hawaiian population). Mottled pattern consists of modest to small dark spots and blotches on neck, back, and sides of trunk; mottling varies from moderately dense to light. Striped pattern is superimposed on mottled background or near uniform ground color. Stripes are moderately broad and light colored (cream to tan); dorsolateral stripes extend from near eye to rear of trunk. Most individuals have dorsal dark bar between the eyes; this mark can join with larger, variably shaped dark mark on rear of head and anterior neck. Dark horizontal stripe is usually present from rear of eye to above tympanum and curving downward to forelimb insertion. Underside is typically white with no dark stippling. Hindlimbs are distinctly barred light and dark. Fore- and hindlimbs are slender and long; fore- and hindfeet are web-free. All digits have large terminal disks whose distal end is more square than rounded.

SIZE: Adult females average larger than males. Adult females attain a maximum length of ~ 36 mm SVL but usually are much smaller, with a typical body range of 19 to 26 mm SVL; males range from 15 to 18 mm SVL.

OCCURRENCE: *E. planirostris* is a widespread species in Cuba and the islands of the Bahaman Bank. It appears to have reached the Florida Keys before the arrival of humans, hence it is considered a native of southern Florida. Humans brought it to the Hawaiian Islands (Oʻahu and Hawaiʻi) in the early 1990s; it was first reported in Guam in 2003.

HABITAT CHOICE: Naturally *E. planirostris* is an inhabitant of moist forest but it has adapted well to human modified habitats that remain moist, such as greenhouses, plant nurseries, and house-side shrubbery that receives frequent watering. These frogs are mainly terrestrial and live amid the surface leaf and other surface litter—including human trash—that remain moist or offer deeper moist retreats during periods of surface drying.

REPRODUCTION: Greenhouse frogs exhibit direct development. Females lay 2 to 22 eggs, (clutches average in the mid-teens), in a moist site under leaf detritus and other surface cover that creates a constantly damp environment. Clutch size has not been determined for the populations of the Hawaiian Islands and Guam. Presumably the nest site is selected by the male, although reportedly he does not remain to guard the eggs from predators and desiccation until they hatch. Froglets hatch in two to four weeks; the speed of development is temperature-dependent. Reproduction occurs year-round where temperature and humidity remain high; however, the seasonality has not been determined for either of the Pacific populations. Calling males produce a short, gentle birdlike chirping trill repeated four to six times, usually from a calling site beneath vegetation.

MISCELLANEA: These frogs consume a variety of small invertebrates.

Hylidae: Treefrogs

Hylids are largely treefrogs with long slender limbs and each digit capped by an enlarged digital pad. A few of the more than 900 hylid species are terrestrial or semiaquatic and have body-plans that are like those of ranid frogs. Most species lay eggs in an aquatic situation; the eggs hatch into free-swimming larvae (tadpoles). As in most speciose frog groups, adult body size ranges from small (20–22 mm SVL) to over 100 mm SVL in a few species. The family has three evolutionary lineages, classified as subfamilies. The Leaf Frogs (Phyllomedusinae) occur in Central and South America. The Australian Treefrogs (Pelodryadinae) consists of a single speciose (~200 species) genus, *Litoria*. *Litoria* is the dominant frog of Australia and New Guinea, comprising about 90 percent of frog species in Australia and roughly one-third of the species in New Guinea, where microhylids and ceratobatrachids also have many species. The most speciose lineage, the Hylinae, is predominantly a New World diverse group. Only *Hyla* occurs in Eurasia and only with a few species spottily distributed across the breadth of this huge landmass.

Litoria aurea   Green and Gold Bellfrog

Plate 2

APPEARANCE: Midsize to large, robust-bodied frog with smooth, glossy skin. Dorsally, *L. aurea* is boldly colored in green and gold between light-colored (white to yellow) dorsolateral stripes that extend from snout nearly to hindlimbs. On trunk, this stripe covers dorsolateral fold on each side. Dorsal ground color is green and amount of gold (to tannish-brown) spotting varies from none to near replacement of green. Dorsolateral stripe is commonly edged below by black. Upper lip is light green to white, and this stripe extends posteriorly below golden tympanum to shoulder. Groin is greenish blue. Ventral surface is granular and white. Fore- and hindlimbs are moderately robust and long. Forefeet are web-free; hindfeet are about two-thirds webbed with web reaching basal joint of terminal or penultimate phalanx of three outer toes. All digits have expanded terminal pads that are round in general outline.

SIZE: Adult females average larger than males. Adult females range in size from 65 to 108 mm SVL; males from 57 to 69 mm SVL.

OCCURRENCE: This treefrog is an alien species in Vanuatu. It probably arrived in the 1960s, presumably from New Caledonia. Its introduction and spread throughout Vanuatu occurred largely through the efforts of plantation owners in anticipation of insect control. Presently, it occurs only on Efate, Espiritu Santo, and Malakula. *L. aurea* is a native of eastern Australia, where it has a modest distribution in the coastal area of New South Wales to eastern Victoria.

HABITAT CHOICE: In its Australian homeland, this treefrog regularly occurs in dense bulrush vegetation around permanent ponds. In Vanuatu, it occupies this type of habitat as well as a variety of dense, scrubby thickets adjacent to pastures in agricultural areas, occasionally distant from permanent water.

REPRODUCTION: External fertilization (amplexus) and indirect development. Reproductive behavior has not been described for the Vanuatu population. In Australia, males call (a slow call beginning with series of "crawks" and then grunts, and ending with a "crok-crok") while floating in the water. Females lay large masses of eggs (4,000–6,000) that adhere to floating and emergent vegetation near the water's edge. Eggs typically hatch in three or four days, and free-swimming larvae (tadpoles) feed for an average of three to four months before metamorphosing into froglets.

Litoria fallax   **Eastern Dwarf Treefrog**
Plate 2

APPEARANCE: Small frog with relatively smooth skin on top and sides of body, underside lightly granular. *L. fallax* is variably colored, ranging from uniform green or medium-brown dorsal ground color to a two-tone dorsal pattern of a light brown to yellow dorsum and green sides. Bright-white lip is almost always bordered above by black from snout to the tympanum. Venter is white from chin to vent, and thighs are orange fore and aft. Fore- and hindlimbs are slender and long; forefeet are barely webbed, and hindfeet are distinctly webbed with the web extending to the basal joint penultimate phalanx of three outer toes. All digits have expanded terminal pads, round in general outline.

SIZE: Adult females average larger than males. Adult females typically ranges from 25 to 32 mm SVL; males from 22 to 26 mm SVL.

OCCURRENCE: First reported in 1968, this treefrog is an alien in Guam. It is a native of eastern Australia, occupying a modest coastal distribution from southwestern Queensland into northeastern New South Wales, Australia. It appears to be widespread, although spottily distributed throughout much of Guam, and is probably established most successfully in the marshy area of the southern half of the island.

HABITAT CHOICE: In its native Australia, this coastal dwarf species lives in vegetation bordering ponds, marshes, and slow-moving streams. It occupies similar habitats in Guam.

REPRODUCTION: External fertilization (amplexus) and indirect development. Females lay eggs singly or in small clusters on plant stems in still-water areas, usually areas that have thick emergent vegetation. The eggs typically hatch in three to five days. Free-swimming larvae (tadpoles) feed for 10 to 20 weeks before metamorphosing into froglets. Males call from perches on emergent vegetation. Their voice is a high pitched "wreek-pip-pip," and sounds like someone running a fingernail over a fine-toothed comb.

MISCELLANEA: The arrival of *L. fallax* in Guam is undocumented, although it was probably introduced in the early 1960s.

Scinax quinquefasciatus   **Five-lined Snouted Treefrog**
Plate 2

APPEARANCE: Small frog with smooth to lightly granular skin dorsally and laterally; ventral skin of chest and abdomen with round, flat tubercles. Mutely colored dorsally and laterally in shades of medium and dark brown. Dorsal pattern typically consists of a large, medium-brown triangular blotch extending from between eyes to nape, broad brown dorsolateral stripes with brown mottling between them, lateral brown stripe on each side from loreal area to midtrunk and usually bordered above and below by narrow, beige stripes. Entire underside from chin to thighs is uniform whitish beige, underside of thighs dusky. Fore- and hindlimbs are slender and long. Forefeet are web-free, hindfeet strongly webbed with web reaching halfway to tips of three outer toes. All digits are strongly expanded, each with oblong pads perpendicular to digit axis.

SIZE: Adult females and males are similarly sized. Adult females and males typically range from 33 to 38 mm SVL (mean 36 mm SVL).

OCCURRENCE: This treefrog is alien to the Galápagos, apparently arriving in 1997 or 1998 during an exceptionally wet El Niño event. Presently, it resides on the islands of Santa Cruz, San Cristóbal, and Isabela; it will likely spread to other islands in the archipelago. It is a native of the Pacific lowlands of Colombia and Ecuador, and has adapted well to disturbed landscapes.

HABITAT CHOICE: *S. quinquefasciata* is a forest-edge species, and now also lives in human-disturbed habitats associated with agriculture. It is a seasonal breeder that responds quickly to heavy showers and breeds in temporary pools, usually those with emergent vegetation that are subject to rapid drying if rains do not occur regularly.

REPRODUCTION: External fertilization (amplexus) and indirect development. Breeding requires standing water; however, temporary rain pools, even small ones, seem adequate. Females lay small clusters of eggs, which typically hatch quickly and presumably grow and metamorphose quickly.

MISCELLANEA: In spite of the common occurrence of this species in the Ecuadorian and Peruvian coastal lowlands, details of its natural history are poorly studied.

As an alien in the Galápagos Islands, it is a threat to the native fauna and should be eradicated.

Microhylidae: Narrow-mouthed Toads

Microhylid frogs are incredibly diverse and speciose (500 species). This diversity includes a range of sizes, from tiny frogs with adult lengths of 12 mm SVL or less to large species that measure well over 100 mm SVL; species that live in arboreal to fossorial habitats; and species with reproductive cycles that range from typical aquatic tadpole larvae to those with terrestrial eggs with direct development. This diversity is divided among 11 subfamilies. The Marbled Pygmy Frog is a member of the subfamily Microhylinae, which has roughly 80 species in nine genera, all living in Asia; nearly half are members of the genus *Microhyla*. *Microhyla* species are all tear-shaped frogs with tiny, pointed heads tapering smoothly into large globular bodies. The tiny head, and hence the small mouth, gives rise to the name "Narrow-mouthed Toads." Microhylines are usually terrestrial to semifossorial; all appear to have a dietary preference for ants and termites. All microhylines lay eggs in water and have a tadpole stage.

Microhyla pulchra   **Marbled Pygmy Frog**

Plate 4

APPEARANCE: Small frog with small, pointy head atop large, globular body. Dorsal skin is lightly rugose dorsally and laterally, becoming smoother and shiny on venter. Coloration is striking. Dorsal ground color is light brown with pattern of multiple shades of darker browns. Broad V of several shades of light to medium brown extends from between eye, with arms extending backward and downward into groin. Rear edge of V is broadly edged in black and numerous, dark-brown-centered ocelli lie between V's arms. Dorsal pattern of the hindlimbs is continuation of back pattern and hides limbs when frog is in its normal sitting posture. Venter is dusky from chin to anterior chest, white thereafter to groin; hindlimbs are orangish. Forelimbs are small and slender, hindlimbs moderately robust and long. Forefeet are web-free; hindfeet about one-third webbed. Digit tips are blunt on both fore- and hindfeet.

SIZE: Adult females average larger than males. Adult females typically range from 27 to 34 mm SVL (mean 30 mm); males from 24 to 27 mm SVL (25 mm).

OCCURRENCE: This species is largely a Chinese and Southeast Asian inhabitant, found from central and southern China to western Thailand. It was first reported in Guam in 2004, and is now well established in the southern third of the island.

HABITAT CHOICE: The Marbled Pygmy Frog appears to be largely a forest-edge or ecotonal species, hence it has adapted well to anthropogenic disturbances that create temporary ponds in small fields, pastures, and other landscapes.

REPRODUCTION: External fertilization (amplexus) and indirect development. Females lay multiple clusters of 10 to 100 eggs in temporary pools and puddles created by heavy rains. Hatching and developmental data are not available, but both must be rapid because of the use of temporary pools. Calling males produce rapid series of metallic notes.

Ranidae: Water Frogs

The Ranidae is not what it used to be! The preceding sentence is poor grammar but accurately describes the situation of this family prior to Frost and associates' 2006 phylogenetic analysis of the Anura and the resulting revamping of frog classification. The huge admixture of frogs that was the worldwide Ranidae reduced to a monophyletic family of 16 genera and about 350 species, still a speciose grouping. Ranids are a cosmopolitan group, absent only from southern South America and most of Australia. Three genera—*Lithobates* (principally North American), *Rana* (Eurasian), and *Hylarana* (African and South Asian)—contain most of the family's species. All ranids lay eggs in aquatic situations and have a tadpole stage.

Glandirana rugosa   **Japanese Wrinkled Frog**

Plate 3.

APPEARANCE: Small frog with strongly tuberculate and rugose skin. Skin is as "warty" in appearance as toad's skin but lacks large circular glandular warts. Instead, skin has dense covering of tiny to small tubercles. These tubercles occur even on face, including eyelids and tympana. Dorsum of neck, trunk, and hindlimbs has numerous rugae. Dorsal ground color of head, body, and limbs is dark brown, with midsize black spots scattered over trunk but nearly invisible owing to darkness of background color. Venter is dusky brown. Rear surface of thighs is dark and light mottled; dark mottling occasionally coalesces into few vertical bars or horizontal stripe. Fore- and hindlimbs are moderately slender and long. Forefeet are web-free; hindfeet are strongly webbed, with web reaching to basal joint of terminal or penultimate phalanx of three outer toes.

SIZE: Adult females average larger than males. Adult female ranges from 44 to 53 mm SVL (mean 50 mm); males from 37 to 46 mm SVL (41 mm). The average sizes of the Hawaiian population appears to be smaller: 46 mm for females and 34 mm for males.

OCCURRENCE: *G. rugosa* is a Japanese species, first introduced into O'ahu in 1895 or 1896. It presently occurs on Hawai'i, Maui, and Kaua'i as well as O'ahu. Its native range is the southern main islands of Japan: Honshu, Shikoku, and Kyushu.

HABITAT CHOICE: In the Hawaiian Islands, *G. rugosa* occupies taro ponds, roadside puddles, and mountain streams. It is seldom found with the Bullfrog because of the Bullfrog's diet, which includes any animal smaller than itself. In mountain streams, it is found regularly in the shallow upper reaches of the streams.

REPRODUCTION: External fertilization (amplexus) and indirect development. Females lay loose masses of eggs, which generally adhere to vegetation, in standing water or areas of low or no water movement in streams. In each breeding cycle, a female lays 800 to 1,300 eggs, although eggs are seldom all laid/deposited in single mass; some clusters are as small as 10 to 30 eggs. Eggs typically hatch in less than a week. The length of the tadpole stage is not reported for Hawaiian populations. In Japan, tadpoles often overwinter, but such a lengthy larval stage seems unlikely for the mild climate of the Hawaiian Islands. The voice of calling males is a short trill.

MISCELLANEA: A recent study of sex chromosome morphology indicated that there are possibly four or more lineages in Japan. Multiple species hiding under a single name is a common feature among Asian frogs. Only further molecular studies will reveal the actual identity of the Hawaiian population and possibly its Japanese source population.

Hylarana guentheri   **Brown and Tan Amoy Frog**
Plate 3

APPEARANCE: Midsize to large frog with smooth skin and well-developed dorsolateral glandular ridges from eye to groin. It is boldly colored, with contrasting dorsal and lateral browns. Dorsum between dorsolateral ridges is light to medium brown from snout to vent; this area becomes medium brown and mottled in older individuals. Each ridge is dark brown and highlighted by narrow longitudinal strip of grayish brown. Sides are dark brown from behind eye to groin. Ventrolaterally, trunk is white with dark-brown spots. Rear of thighs is heavily mottled brown on white. Underside is white from chin to vent. Fore- and hindlimbs are moderately robust and long. Forefeet are web-free; hindfeet are strongly webbed, with web nearly reaching tips of third and fifth toes. Tips of fingers are bluntly rounded, not expanded; toe tips are slightly expanded and spade-shaped.

SIZE: Adult females average slightly larger than males. Adult females range from 64 to 80 mm SVL; males from 64 to 75 mm SVL.

OCCURRENCE: *H. guentheri* is a recent arrival (late 1990s or early 2000s) to Guam from Asia, likely from Taiwan, and now widespread and abundant throughout Guam. This species occurs broadly from central Vietnam throughout southern China to Yangtze River valley and on the islands of Hainan and Taiwan.

HABITAT CHOICE: The Brown and Tan Amoy Frog is a lowland inhabitant of slow-moving permanent water, living in a variety of habitats from ponds, marshes, and small streams. It has adapted well to agricultural modification and thrives in paddies and irrigation canals.

REPRODUCTION: External fertilization (amplexus) and indirect development. Females generally lay large clusters of several thousand eggs in areas of emergent vegetation in standing water or streams with low or no water movement. Data on hatching and developmental time are not available. Calling males produce two- to four-note calls of "chuuk – chuk"; the male's voice has also been likened to a barking dog.

MISCELLANEA: This frog presently appears more abundant in Guam than the Cane Toad. Its dispersal throughout Guam results, at least in part, from its ability and tendency to travel long distances from its waterside residence.

Lithobates catesbeianus   **American Bullfrog**

Plate 3

APPEARANCE: Large, heavy-bodied frog with smooth skin. Aside from distinct supratympanic ridge, only a few low glandular tubercles are present dorsally on posterior quarter of trunk. Its ground color ranges from green to olive brown. Darker spotting or mottling is commonly evident on body and limbs, although dorsum may be unicolor. Venter has cream to light-tan ground color and is strongly mottled with brown to olive, thick reticulations. Fore- and hindlimbs are robust and long. Forefeet are web-free; hindfeet are strongly webbed with the web extending nearly to tips of the third and fifth toes. All digits have narrow to blunt tips.

SIZE: Adult females average larger than males. The body size of mature bullfrogs varies geographically, and data for the Hawaiian populations are not available. In cool temperate native localities, mature individuals range from 85 to 105 mm SVL; in warmer native localities, maturity does not occur until it reaches 120 mm or more. Japanese *L. catesbeianus* females average 162 mm SVL (120–183 mm); males average 152 mm (111–178 mm). A maximum length of 202 mm SVL has been reported for a North American female.

OCCURRENCE: Bullfrogs are native to eastern North America from the high-grass prairie eastward, extending from cold-temperate Nova Scotia to subtropical mid-peninsular Florida. Because of their possibility as a commercial food source, they have been introduced widely throughout the world. Only Australia and Africa appear to have escaped introductions. In the tropical Pacific, these frogs are well established on several Hawaiian Islands. The original Hawaiian introduction was pre-1900s, with individuals derived from populations originally introduced to California.

HABITAT CHOICE: Bullfrogs are aquatic and rarely found far from permanent water. They prefer heavily vegetated shorelines of streams, ponds, lakes, marshes, and swamps.

REPRODUCTION: External fertilization (amplexus) and indirect development. Males call in small groups away from the shore, generally amid floating or low-emergent vegetation. Each female lays thousands of eggs in a thin surface film. Eggs typically hatch in less than a week, and free-swimming larvae (tadpoles) feed for two months or longer before metamorphosing into large froglets. In localities with cold winters, tadpoles can require one or two years to grow to metamorphosis size. Hawaiian tadpoles grow rapidly and metamorphose in about six months. The male's call is aptly described as a bass "jug-of-rum."

MISCELLANEA: Bullfrogs are aggressive and opportunistic carnivores. They readily eat any animal that they can capture, including smaller Bullfrogs. This behavior makes them a dangerous invader and results in their decimation of the smaller native waterside fauna. In Hawaii, they prey on the young of endangered waterbirds.

Rhacophoridae: Afro-Asian Treefrogs

The Rhacophoridae is a moderate diverse group of two subfamilies, Buergeriinae and Rhacophorinae. All family members possess a treefrog body form, and their hands and feet have expanded digit tips; however, not all species are arboreal. The buergeriines consist of four species with semiaquatic habitats. This subfamily occurs only in

Taiwan and southern Japan. The rhacophorines are much more diverse, presently consisting of 13 genera and about 320 species. They are also geographically widespread; members occur in sub-Saharan Africa and tropical Asia from Sri Lanka and India to Taiwan and the Philippines. Body sizes range from 20–22 mm to 100–110 mm SVL. Reproduction includes species with direct development and others with indirect development. The indirect-developing species create foam nests suspended over water. Once the larvae hatch, they wiggle through the foam and drop into the water. The foam nest is created from the female's oviducal secretion; as the eggs are laid, the amplexing pair stir the secretion with their hindlimbs, producing the meringue-like foam. This manner of foam creation gave rise to the name "Whipping Frogs."

Polypedates braueri   **Taiwan Whipping Frog**

Plate 2

APPEARANCE: Midsize frog with smooth to lightly granular skin dorsally and laterally; ventral skin of chest, belly, and thighs is distinctly granular with round, flat tubercles. This treefrog is mutely colored in shades of light to medium brown dorsally and dorsolaterally, and varies from unicolor to modestly spotted in irregularly shaped, dark spots, often forming X-like pattern on shoulders. Typically, its face from snout to ear has dark-brown mask contrasting with lighter-colored upper lip. Anterodorsal surface of thigh has narrow brown bars on light-brown background; rear of thigh is medium brown with more or less double row of white spots on dark-brown background. Ventral surface is light tan. Fore- and hindlimbs are slender and long. Forefeet are web-free; hindfeet are strongly webbed with web reaching basal joint of terminal phalanges of three outer toes. All digit tips are strongly expanded and circular, distal edge is slightly truncate.

SIZE: Adult females are larger than males. Adult females range from 60 to 75 mm SVL; males from 45 to 55 mm SVL.

OCCURRENCE: The Whipping Frog is a recent arrival in Guam, first reported in 2004. It now has established populations in the Santa Rita, Yona, Inarajan, and Naval Magazine areas. It appears to be slowly expanding from these areas. It is a native of Taiwan and eastern China.

HABITAT CHOICE: In its homeland, *P. braueri* lives in a variety of agricultural landscapes—particularly those that have scrubby fencerows or small patches of forest nearby. It readily breeds in paddies and other temporary ponds, preferring those with dense vegetation in shallow water.

REPRODUCTION: External fertilization (amplexus) and indirect development. Whipping Frogs derive their name from the leg-beating of the egg mass by the male and female as the female extrudes the eggs from her vent. For *P. braueri*, the foam nests of 300 to 400 eggs are attached to the stalks of emergent vegetation in shallow water. Eggs usually hatch in few days; the larva writhes free of the foam and drops into water. The free-swimming larvae (tadpoles) feed for about eight weeks before metamorphosing into froglets. Calling males produce a short series of grunts or snore-like vocalizations.

MISCELLANEA: *P. braueri* was only recently (2011) separated from *P. megacephalus*. The identity of the Whipping Frog in Guam is tentative, although Guam individuals most closely match the traits offered in the resurrection of *P. braueri*.

LIZARDS

Lizards are the dominant terrestrial reptile of the tropical Pacific. If an island is large and high enough to support a reptile, that reptile will be a species of lizard. Lizards have proved to be the most successful island colonizers among terrestrial vertebrates, excluding birds that fly easily across saltwater barriers. Reptiles are characterized by cornified (keratinized) skin; in lizards, the skin surface is composed of scales. These scales range from small abutting granules, blocks, and tubercles to large, shiny, overlapping scales, and they occur in a multitude of shapes, sizes, and surfaces. This keratinized skin cover makes lizards nearly impermeable to water loss or gain across the skin surface, giving them the ability to withstand immersion in saltwater. Although it is unlikely that any lizard swam from one distant island to another, they can and did traverse great distances by riding rafts of vegetation or hiding in semi-rotten logs. Clutches of eggs embedded within rafts of vegetation provide another likely transit mode. Eggs of gecko species have been shown experimentally to tolerate hours of immersion in saltwater, so they could withstand periodic immersion during stormy seas.

The champion long-distant traveler had to be the ancestor(s) of the Fijian iguanas. The closest relatives of these lizards are the iguanas of the tropical America. The Americas are thousands of miles from the Tongan and Fijian archipelagos; nevertheless, they are downstream of South America via the South Equatorial Current. Phylogenetic relationships and a likely transport mechanism are strong evidence that support the ancestoral origins of the Southwest Pacific iguanas.

GEKKOTA: GECKOS

In the tropics, no lizard is more familiar to us than the geckos. They cling to the walls of homes and restaurants, catching the moths and other insects attracted to our lights. These house geckos, however, are a small fraction of the diversity of the Gekkota, usually although not exclusively members of the genus *Hemidactylus*, a gekkonid lizard. The Gekkota consist of seven families: Carphodactylidae, Diplodactylidae, Pygopodidae, Eublepharidae, Sphaerodactylidae, Phyllodactylidae, and Gekkonidae. The first three families are Southwest Pacific and Australian groups. Considerable diversity is found within these three closely related families, where species range from tiny to giant geckos, terrestrial to arboreal species, and even the legless Pygopodidae. The other four families represent another evolutionary group (clade). The eublepharid geckos are the earliest divergence of terrestrial geckos, and now live in the North American Southwest, Africa, and southern Asia. The latter three families have representatives in the Pacific: Sphaerodactylidae with a single Pacific species on Isla de Coco, Phyllodactylidae with a small species radiation on the Galápagos Islands, and the Gekkonidae with numerous genera and species in Oceania. All three families are speciose and occur broadly; the former two are principally tropical American and African families, and the Gekkonidae occur pantropically, although with its greatest diversity in Africa and Asia.

Gekkonidae: Geckos

The pantropic Gekkonidae is the most speciose family of geckos, with more than 40 genera and nearly 900 species. These geckos range in body size from small to large, although most species lie within body size range of 35 to 150 mm SVL. There are numerous terrestrial species with narrow elongate digits and many species with digit tips modified in diverse ways to provide climbing pads for a fully arboreal existence. Pacific geckos are predominantly in this family.

Gehyra brevipalmata   **Palau Gecko**

Plate 5

APPEARANCE: Midsize, stout-bodied lizard with short, robust limbs. Head, body, and tail are light to medium brown to olive above with numerous irregular, small white spots and few to numerous rufous spots on neck and trunk. Venter is uniformly bright to pale yellow. Small, equal-sized granular, juxtaposed scales cover head, neck, and trunk dorsally and laterally; scales flatten ventrolaterally and are larger and imbricate ventrally. Head has small, smooth plate-like scales bordering mouth and nares; elsewhere granular scales cover head; chin bears single series of postmental scales. Tail is thick, subcylindrical, and gradually tapers to blunt tip; it has granular scales above, grading into small flat, slightly imbricate scales ventrolaterally, and row of transversely enlarged smooth scales midventrally. Tail lacks segments. There are 11 to 13 subdigital lamellae on fourth toe of hindfoot. Adult males have 30 to 46 precloacal-femoral pores in continuous series.

SIZE: Adults of this gecko are sexually dimorphic. Adult females range from 62 to 77 mm SVL (mean 71 mm); males from 66 to 89 mm SVL

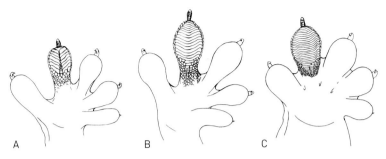

FIGURE 3. Digital pads of Pacific *Gehyra*. Ventral view of the hindfeet of (A) *G. insulensis*, (B) *G. oceanica*, and (C) *G. vorax*. The morphology of the lamellae beneath the digits is helpful in differentiating among species in many lizards. For these three geckos, the shape of the pad also differs, as does the number and division of the subdigital lamellae.

(77 mm). Tail length is 90% to 100% of SVL. Hatchlings are about 24 to 26 mm SVL.

OCCURRENCE: *G. brevipalmata* is a Palauan endemic. It occurs widely throughout the Palauan Archipelago.

HABITAT CHOICE: The Palau Gecko was originally a forest dweller and is still common within the forest; however, it has also adapted to human disturbance and lives in gardens and on buildings. When resting, it is a crevice dweller, occupying any tight space, such as a rafter immediately beneath a roof in a building or natural crevices in pandanus and bamboo axils and trunk crevices of banyan trees.

REPRODUCTION: Detailed reproductive data are not available. It typically lays two eggs in crevices.

MISCELLANEA: It is nectarivorous and eats fruit in captivity; nonetheless, it probably is mainly insectivorous.

Gehyra georgpotthasti   **Vanuatu Giant Gecko**

Not illustrated

APPEARANCE: Midsize to large, stout-bodied lizard with short, robust limbs. Head, trunk, and tail are commonly mottled in large blotches of light to medium brown, olive, and gray. Skin tone can darken and lighten, intensifying contrast between multicolored blotches or becoming nearly unicolor. Venter is light brown with yellow tint. Small, equal-sized granular, juxtaposed scales cover head, neck, and trunk dorsally and laterally; scales flatten ventrolaterally and are larger and imbricate ventrally. Head has small, smooth plate-like scales bordering mouth and nares; elsewhere granular scales cover head; chin bears single series of postmental scales. Tail is thick, subcylindrical, and gradually tapers to blunt tip; it has granular scales above, grading into small flat, slightly imbricate scales ventrolaterally, and row of transversely enlarged smooth scales midventrally. Tail encircled by alternating rows of enlarged scales. There are 21 to 30 subdigital lamellae on fourth toe of hindfoot. Adult males have 27 to 68 precloacal-femoral pores in continuous series.

SIZE: Adults are sexually dimorphic. Adult females range from 90 to 125 mm SVL (mean 105 mm); males from 82 mm to 142 mm SVL (115 mm). Tail length is 70% to 80% of SVL. Hatchlings size has not been reported.

OCCURRENCE: *G. georgpotthasti* has just been recognized as a distinct species within the *G. vorax* complex. It occurs widely throughout Vanuatu and in the Loyalty Islands. Two individuals were found recently on Tuamotu (Fakarava), and there are 19th-century specimens from the Society Islands. These Polynesian outliers are most likely the result of human transport.

HABITAT CHOICE: This gecko lives in the canopy of trees, mainly in moist evergreen forest and late-stage secondary forest, but it also lives in banana plantations and coastal localities with only pandanus and coconut palms.

REPRODUCTION: Females lay two large, spherical eggs, presumably in tree holes, beneath bark, and in palm axils. Other reproductive data are lacking.

MISCELLANEA: It is vocal with a bark like a small dog and may be mistaken for a fruit dove.

Gehyra insulensis   **Pacific Stump-toed Gecko**
Plate 5, Figure 3

APPEARANCE: Midsize, stout-bodied lizard with short, robust limbs. Head, body, and tail are light to medium grayish brown above with numerous irregular dark markings on head to and onto tail. Skin tone can darken and lighten and obscure the markings. Venter is uniformly grayish white to light yellow. Small, equal-sized granular, juxtaposed scales cover head, neck, and trunk dorsally and laterally; scales flatten ventrolaterally and are larger and imbricate ventrally. Head has small, smooth plate-like scales bordering mouth and nares; elsewhere granular scales cover head; chin bears single series of postmental scales, midline pair largest. Tail is subcylindrical to dorsoventrally flattened and gradually tapers to blunt tip; dorsal surface uniformly covered with granular scales, ventrolateral edge with frayed appearance created by two rows of small, narrow, elongate scales; ventrally small flat, slightly imbricate scales with median row of transversely enlarged smooth scales. Tail shows faint segmentation. There are eight to nine subdigital lamellae (all medially divided) on fourth toe of hindfoot. Adult males have 32 to 40 precloacal-femoral pores in continuous series.

SIZE: Adult females and males are equal in size. Adult females range from 42 to 50 mm SVL; males from 36 to 50 mm SVL. Tail length is 100% to 110% of SVL. Hatchlings are 18 to 23 mm SVL.

OCCURRENCE: *G. insulensis* is widespread throughout the Pacific, although uncommon in most islands.

HABITAT CHOICE: Although the Pacific Stump-toed Gecko is considered a recent arrival in Oceania, it is an uncommon house gecko. When living on buildings, it usually occupies the unlighted sides. More commonly, it occupies semi-natural habitats from house-side and garden shrubbery to secondary and undisturbed forest. Rarely, except in the Hawaiian Islands, is it a resident of primary forest; there it resides also on exposed lava outcrops.

REPRODUCTION: Females typically lay two eggs, rarely one, in cracks of buildings, beneath bark, and in palm axils and other narrow spaces. Egg-laying probably occurs year-round, although that inference has not been confirmed for tropical Pacific populations. The eggs hatch in about 60 to 65 days.

MISCELLANEA: *G. insulensis* is strongly nocturnal. During the day, individuals rest in arboreal crevices such as spaces beneath bark, in tree holes, or within clusters of epiphytes. At night, it appears to adopt a sit-and-wait strategy, making quick, short dashes to capture insect prey.

Among Pacific geckos, a skin-shedding escape mechanism is particularly evident in *G. insulensis* and *G. oceanica*. The upper layer of the skin of these geckos is loosely connected to the lower layer. When grasped by a predator, they twist their body or limb and the predator is left holding a piece of skin, allowing the gecko to escape.

The name *insulensis* replaces *mutilatus* for Pacific populations. A recent molecular study demonstrated that populations formerly considered *G. mutilata* consisted of two genetically distinct lineages, a southern Asian and Indian Ocean one and an Oceania and Southwest Pacific one. Although the two species show slight morphological differentiation, the recent molecular data confirm an earlier hypothesis of genetic divergence, hence the adoption of the name *G. insulensis* here.

Gehyra oceanica   Oceania Gecko
Plate 5, Figure 3

APPEARANCE: Midsize, stout-bodied lizard with short, robust limbs. Head and trunk are usually unicolor light to dark olive-brown or gray, tail often with alternating bands of light and dark. Skin tone can darken and lighten. Commonly, a pale stripe extends from snout through eye to above ear-opening. Venter is either uniformly grayish olive or yellow. Small, equal-sized granular, juxtaposed scales cover head, neck, and trunk dorsally and laterally; scales flatten ventrolaterally and are larger and imbricate ventrally. Head has small, smooth plate-like scales bordering mouth and nares; elsewhere granular scales cover head; chin bears single series of postmental scales, midline pair largest. Tail is thick, slightly flattened cylinder and gradually tapers to blunt tip; it has granular scales above, grading into small flat, slightly imbricate scales ventrolaterally, and row of transversely enlarged smooth scales midventrally. Tail is distinctly segmented; each segment has one or two slightly enlarged scales ventrolaterally near its posterior edge. There are 12 to 16 subdigital lamellae on fourth toe of hindfoot. Adult males have 17 to 41 (usually less than 30) precloacal-femoral pores in continuous series.
SIZE: Adult females average slightly larger than males. Adult females range from 64 to 84 mm SVL; males from 59 to 82 mm SVL. Tail length is about the same as SVL. Hatchlings range from 28 to 34 mm SVL.
OCCURRENCE: *G. oceanica* occurs widely throughout the central and western Pacific islands. It is absent from the eastern Pacific islands and the Hawaiian Islands.
HABITAT CHOICE: The Oceania Gecko was probably originally a forest and forest-edge inhabitant. Now it occupies the full range of habitats, from buildings and gardens to the interior of forests. It appears most abundant in rural habitats from garden edge to early secondary-growth forest. It also basks in the shade during the day, especially in afternoons.
REPRODUCTION: Females appear to lay eggs year-round. Clutch size is regularly two eggs, occasionally one, laid in arboreal crevices. Finding a single clutch of eggs is rare because *G. oceanica* is strongly a com-

munal nester with multiple females using the same crevice or hole and repeatedly returning to the same nesting site. These communal nests can have more than a hundred eggs in all stages of development and hatched shells. Incubation is long, with recently deposited eggs requiring over 100 days (102–114) to hatch.

MISCELLANEA: *G. oceanica* is an opportunistic carnivore. Its diet is predominantly insects and other invertebrates, but, owing to its size, it preys on other geckos, including its own juveniles.

Crombie and Pregill noted in their review of the Palauan herpetofauna that the large *Gehyra* from the Southwest Islands had some similarities to *G. oceanica*, but it also possessed a number of differences suggesting that it may represent a distinct species. True or typical *G. oceanica* occurs throughout the main islands of Palau. Similarly, some morphological and genetic data suggest the presence of north and south Pacific species.

Gehyra vorax   Fiji Giant Gecko

Plate 5, Figure 3

APPEARANCE: Midsize to large, stout-bodied lizard with short, robust limbs. Head, trunk, and tail are commonly mottled in large blotches of light to medium brown, olive, and gray. Skin tone can darken and lighten, intensifying contrast between multicolored blotches or becoming nearly unicolor. Venter is white, occasionally with some dark mottling. Small, equal-sized granular, juxtaposed scales cover head, neck, and trunk dorsally and laterally; scales flatten ventrolaterally and are larger and imbricate ventrally. Head has small, smooth plate-like scales bordering mouth and nares; elsewhere granular scales cover head; chin bears single series of postmental scales. Tail is thick, subcylindrical, and gradually tapers to blunt tip; it has granular scales above, grading into small flat, slightly imbricate scales ventrolaterally, and has row of transversely enlarged smooth scales midventrally. Tail encircled by alternating rows of enlarged scales. There are 23 to 34 subdigital lamellae on fourth toe of hindfoot. Adult males have 26 to 90 precloacal-femoral pores in continuous series.

SIZE: Adults may be sexually dimorphic. The largest known individual was a 188 mm SVL female; range and mean are not available for females. Adult males range from 90 mm to 156 mm SVL (mean 123 mm). Tail length is 50% to 70% of SVL. Hatchling size has not been reported.

OCCURRENCE: *G. vorax* occurs on the larger islands of Fiji. A Tongan occurrence is based on a single museum specimen and requires confirmation by new sightings.

HABITAT CHOICE: This gecko lives in the canopy of trees, mainly in primary evergreen forest and late-stage secondary forest. In rural areas with nearby fragmented forests, it is seen in bread fruit trees and palms.

REPRODUCTION: Females lay two large, spherical eggs, presumably in tree holes, beneath bark, and in palm axils. Other reproductive data are lacking.

MISCELLANEA: *G. vorax* is reported to be a fruit eater, and it survives on a fruit diet in captivity. However, it is likely that free-living individuals also eat invertebrates and small vertebrates. Observations indicate that in some areas it is active during the day as well as at night.

Gekko gecko   **Tokay**

Plate 6, Figure 4

APPEARANCE: Large, moderately robust lizard with moderately short, robust limbs. Head, body and tail have bright orange spots on pale-green through grayish-olive to brownish or bluish-gray background. Juveniles and subadults have transverse rows of white spots on neck, trunk, and limbs. Venter is uniformly light yellow to cream. Dorsally and laterally body is covered with small, juxtaposed scales, granular on top and side of head, flattened on trunk with rows of tubercles; larger, smooth, slightly imbricate scales anteriorly on limbs and ventrally from midthroat onto tail. Dorsally, trunk bears 10 to 14 longitudinal rows of flattened round and trihedral tubercles. Hindlimbs have few to no tubercles on their dorsal surface. Head has small, smooth, plate-like scales bordering mouth and nares, and enlarged granular scales on snout. Tail is lightly tapered cylinder with flattened granular scales above, grading into small, flat, slightly imbricate scales ventrolaterally, and row of enlarged smooth scales midventrally. Tail is segmented, and each segment has in its middle a whorl of pointed tubercles. Venter has 135 to 142 ventral scale rows from postmentals to vent, 81 to 105 scales around midbody, and 17 to 24 subdigital lamellae on fourth toe of hindfoot. Adult males have 10 to 18 precloacal pores and no femoral pores.

SIZE: Adults may be sexually dimorphic, but no one has statistically validated dimorphism. Additionally, there appears to be a geographic difference in body size. Females attain maturity at ~115 mm SVL, males

FIGURE 4. Chin scales of Pacific *Gekko*. (A) Palau species. (B) *G. vittatus*. (C) *G. gecko*. The arrangement and relative size of the postmental scales behind the medial mental, and the chin scales abutting the labial scales edging the mouth, discriminate these three Pacific *Gekko*. Note the uniformity of postmental and chin scales in *G. gecko* in comparison with those of the two native Pacific *Gekko*.

at ~135 mm SVL in Bali; 120 mm and 134 mm SVL, respectively, in Komodo. Philippine adult *G. geckos* are 122 to 140 mm SVL (females) and 117 to 155 mm SVL (males). Tail length is 90% to 110% of SVL. Philippine hatchlings are 40 to 42 mm SVL.

OCCURRENCE: Although the Tokay is now well established in southern Florida, its presence as reproducing populations on Guam and the Hawaiian Islands (Oʻahu) remains uncertain. The regular nocturnal surveys for Brown Treesnakes on Guam have not seen or heard this gecko recently. On Oʻahu, the Manoa Valley population apparently was eradicated in the late 2000s, but Tokays may persist elsewhere or subsequent illegal releases by hobbyists may have established new populations. *Gekko gecko* occurs widely throughout southern Asia from eastern India through Indochina to southwestern China, southward through Indonesia to Sulawesi and Timor and eastward into the Philippines; however, there are unexplained absences, such as most of mainland Malaysia, within this broad range.

HABITAT CHOICE: The Tokay is a forest gecko that has adapted well to anthropogenic habitats, including village gardens and buildings as well as suburban-urban horticultural landscapes.

REPRODUCTION: Female Tokays regularly lay two eggs, uncommonly one or three spherical, hard-shelled, adherent eggs. Females deposit eggs in crevices and leaf axils, often repeatedly in the same location; sometimes the deposition site is shared with other females in communal nesting. Reported incubation ranges from 46 to 80 days, dependent upon the ambient seasonal and regional temperature—that is, cooler temperatures result in longer incubations. Similarly, where temperature and food resources permit, females are reproductively active year-round.

MISCELLANEA: Tokays are opportunistic carnivores, feeding predominantly on insects, although they capture a variety of small vertebrate prey, such as geckos and mice. They are nocturnal predators and predominantly sit-and-wait ones. They actively forage when the abundance of their usual insect prey is low.

Gekko vittatus   **Melanesia Ghost Gecko**

Plate 6, Figure 4

APPEARANCE: Midsize to large, slender lizard with short, slender limbs. Dorsal ground color ranges from light tan to medium rufous brown, although often unicolor, anteriorly forked middorsal cream stripe extends from nape onto tail; tail ground color regularly darker than trunk and white banded. Head is grayish brown, occasionally with darker-brown reticulations above; laterally broad dark-brown stripe extends from snout through eye, above ear-opening to midneck. Venter is uniformly light yellow to cream, occasionally grayish. Granular, juxtaposed scales cover body surfacedorsally and laterally; smooth, imbricate scales cover limbs anteriorly and venter from midthroat onto tail. Additionally, trunk has 20 to 30 longitudinal rows of weakly trihedral and cone-shaped tubercles. Fore- and hindlimbs also densely covered with pointed tubercles. Head has smooth, plate-like scales bordering mouth and nares, enlarged granular scales on snout, and mixture of flattened granular scales and tubercles on remainder of head. Tail is lightly tapered cylinder with flattened granular scales above, grading laterally and ventrally into small, smooth, flat, slightly imbricate scales. Tail is segmented, and each segment has two to three whorls of pointed tubercles, largest tubercles on the posterior edge of each segment. Venter has 176 to 222 ventral scale rows from the postmentals to vent, 129 to 155 scales around midbody, and 18 to 26 subdigital lamellae on fourth toe of hindfoot. Adult males have 49 to 65 precloacal-femoral pores.

SIZE: Adults appear to be sexually dimorphic; females range from 84 to 124 mm SVL (mean, ~99 mm SVL) and males from 88 to 119 mm SVL (~106 mm SVL). Tail length is 90% to 118% of SVL and prehensile.

OCCURRENCE: *G. vittatus* occurs from Seram and Halmahera eastward across northern New Guinea into the Solomon Islands and to the Torres and Bank Islands of Vanuatu. Two outlier populations exist, one throughout Timor and another in a small area of Java. If its current presence can be confirmed, the Javan population is probably an introduction.

HABITAT CHOICE: This species is a forest and forest-margin dweller that has adapted well to human gardens and rural settlements. It is strongly arboreal and nocturnal, spending its days resting both openly on tree trunks and vines and hidden in the crevices of tree trunks and axils of palms, bananas, and pandanus.

REPRODUCTION: Females lay two adherent eggs in crevices, commonly using the axils and dead leaf masses of pandanus. Often these sites are shared by multiple females. Communal nests will have both hatched eggshells and incubating eggs, ranging from a dozen to hundreds of eggs and shell fragments. Incubation data are unavailable, although 50 to 60 days is likely in most lowland areas.

MISCELLANEA: *G. vittatus* is an active nocturnal predator, moving across branches and vines, often assisted by its prehensile tail. Its main prey consists of insects and spiders, and occasionally a smaller gecko.

Gekko nsp   **Palau Ghost Gecko**

Plate 6, Figure 4

APPEARANCE: Large, slender lizard with short, slender limbs. Dorsal ground color ranges from gray to bluish or greenish yellow; often anteriorly forked middorsal cream, irregularly edged stripe extends from nape onto tail; tail ground color is variable, usually similar to trunk with mottled light and dark bands. Head is grayish brown with darker brown reticulations above; laterally dark-brown stripe extends from snout through eye, above ear-opening to midneck. Venter is uniformly light yellow to cream. Granular, juxtaposed scales cover body dorsally and laterally; smooth, imbricate scales cover limbs anteriorly and venter from midthroat onto tail. Additionally, trunk has 22 to 26 longitudinal rows of weakly trihedral and cone-shaped tubercles. Fore- and hindlimbs also densely covered with pointed tubercles. Head has smooth plate-like scales bordering mouth and nares, enlarged granular scales on snout, and mixture of flattened granular scales and tubercles on remainder of head. Tail is lightly tapered cylinder with flattened granular scales above, grading laterally and ventrally into small, smooth, flat, slightly imbricate scales, enlarged midventrally. Tail is segmented, and each segment has two to three whorls of pointed tubercles, largest ones on posterior edge of segment. There are 85 to 101 ventral scale rows from base of neck to vent, 110 to 156 scales around midbody, and 20 to 25 subdigital lamellae on fourth toe of hindfoot. Adult males have 23 to 57 precloacal-femoral pores (mean, 36).

SIZE: Adult females and males are equal in size. Adult females range from 104 to 116 mm SVL (mean, 112 mm), males from 102 to 121 mm SVL (113 mm). Tail length varies from about 90% to 110% of SVL. Hatchling size is uncertain, although individuals of 38 to 42 mm SVL retain a yolk-sac scar.

OCCURRENCE: This gecko is a Palauan endemic and occurs broadly throughout the Palau archipelago.

HABITAT CHOICE: Although it is found on human structures—but always in the darkest areas—the Palau Ghost Gecko prefers habitats with leafy vegetation from garden banana patches to forests. It appears to require tight crevices from leaf axils to tree bark cracks for diurnal resting.

REPRODUCTION: This species, like most geckos, lays two eggs. The hard-shelled eggs adhere to one another. Eggs are deposited in crevices, especially the axils of pandanus, bananas, and palms. The reported 90-day incubation period probably results from rearing in captivity at cool temperatures.

MISCELLANEA: The Palau Ghost Gecko is an opportunistic carnivore that eats a variety of invertebrate and vertebrate prey. It is a nocturnal predator relying mainly on a sit-and-wait strategy, resting on tree trunk, branches, and fronds. In captivity, and probably also in the wild, it eats fruit. Males appear to be strongly territorial.

Hemidactylus frenatus   **Common House Gecko**

Plate 7, Figure 5

APPEARANCE: Midsize, moderately robust, slightly flattened lizard with well-developed limbs. Head, body, and tail have light to medium-gray to dusky-brown ground color above; lateral brown stripe of variable intensity from snout to hindlimb is often absent, but when present is regularly bordered above by white stripe from snout onto neck. Dorsal ground color can change from light to dark. Venter is uniformly whitish to light beige. Small, equal-sized granular, juxtaposed scales cover head, neck, and trunk dorsally and laterally; scales flatten ventrolaterally and are larger and imbricate ventrally. Trunk bears several longitudinal rows of widely spaced, flattened, round tubercles. Head has small, smooth, plate-like scales bordering mouth and nares; elsewhere granular scales cover head; chin bears single pair of large postmental scales. Tail is thick, subcylindrical, and gradually tapers to blunt tip; it has granular scales above, grading into small flat, slightly imbricate

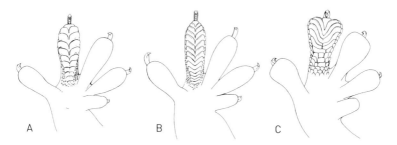

FIGURE 5. Lamellar morphology of the hindfeet of Pacific *Hemidactylus* and *Hemiphyllodactylus*. (A) *Hemidactylus frenatus*. (B) *Hemidactylus garnotii*. (C) *Hemiphyllodactylus typus*. The digital pads of *Hemidactylus* are elongate rectangular, and in *Hemiphyllodactylus*, the pads are circular; the lamellae are U-shaped in the latter and straight in the former.

scales ventrolaterally and midventrally row of transversely enlarged smooth scales. Tail is strongly segmented, each segment posteriorly edged with whorl of five to six large, flattened, cone-shaped scales. There are eight to 11 medially divided, subdigital lamellae on fourth toe of hindfoot. Adult males have 15 to 34 precloacal-femoral pores in continuous series, occasionally small gap in middle of pelvis.

SIZE: Adults of this gecko are sexually dimorphic. Adult females range from 42 to 49 mm SVL (mean 46 mm); males from 48 to 60 mm SVL (51 mm). Tail length is about 100% of SVL. Hatchlings are 19 to 26 mm SVL.

OCCURRENCE: *H. frenatus* has become widespread in the tropical Pacific since the 1970s and was first reported in the Galápagos (Isla Isabela) only in 2011. This gecko was probably a natural resident of Southeast Asia or India, quickly adapted to living with humans, and was transported readily through trade routes. No one has attempted to map the spread of this species, but it was apparently widespread in mainland South Asia and the Sunda, Philippine islands and Guam by end of the 19th century. The widespread transport of military supplies during World War II appears to have established populations in the western Pacific islands, from whence it has spread eastward.

HABITAT CHOICE: The Common House Gecko rarely lives away from human structures and is not common on horticultural vegetation adjacent to buildings. On very small, dry islands, it lives beneath rocks and on trees and shrubs.

REPRODUCTION: Females lay two eggs in crevices (rarely one egg), mostly in manmade structures. Egg-laying occurs in all months of the year.

Eggs hatch in about 70 to 90 days (Samoa, Philippines) and 53 to 54 days (Guam).

MISCELLANEA: *H. frenatus* is strongly nocturnal and usually emerges from the cracks and crevices of buildings only when fully dark. It is aggressive toward other members of its own species and any other gecko, dominating the well-lighted area around lights that attract its insect prey. It will also capture and eat smaller geckos.

Hemidactylus garnotii   Fox Gecko
Plate 7, Figure 5

APPEARANCE: Midsize, moderately slender, slightly flattened lizard with well-developed limbs. Head, body, and tail are light to medium gray to brown above with four longitudinal series of small light-colored spots from neck to base of tail. Dorsal ground color can change from light to dark. Venter is white to light yellow and frequently orangish to salmon on underside of tail. Small, equal-sized granular, juxtaposed scales cover head, neck, and trunk dorsally and laterally; scales flatten ventrolaterally and are larger and imbricate ventrally. Trunk bears no tubercles. Head has small, smooth, plate-like scales bordering mouth and nares; elsewhere granular scales cover head; chin bears single pair of large postmental scales. Tail is thick and distinctly flattened, particularly laterally, and often deltoid leaf-shaped; it has granular scales above, becoming elongate on lateral edge with four projecting cone-shaped scales in each segment, and ventrally larger, smooth, slightly imbricate scales with midventral row of transversely enlarged smooth scales. There are 10 to 15 subdigital lamellae on fourth toe of hindfoot, distal three to four undivided and remainder medially divided.

SIZE: Adults of all-female species gecko range from 49 to 66 mm SVL (mean 56 mm). Tail length is 90% to 100% of SVL. Hatchlings are 20 to 28 mm SVL.

OCCURRENCE: *H. garnotii* is widespread throughout the Pacific, but nowhere is it abundant. It is considered an alien species derived from a Philippine or Sundan population, and it probably arrived in the major Pacific trading ports in the 18th century or earlier.

HABITAT CHOICE: It is not a common house gecko in the Pacific. When found on buildings, it occupies the unlighted walls. It occurs in a variety of modified habitats from gardens to scrubby secondary forest.

REPRODUCTION: All known populations of *H. garnotii* contain only females. This parthenogenetic gecko lays two eggs in cracks in buildings, in leaf axils, and in similar protected sites. Incubation is about 60 to 65 days.

MISCELLANEA: The Fox Gecko is arboreal and a nocturnal predator of insects and other invertebrate prey. Its numbers have greatly declined in the Pacific as *H. frenatus* has spread.

Hemiphyllodactylu ganoklonis　　　　　　　　　**Palau Slender Gecko**

Plate 7

APPEARANCE: Small, slender, elongate-bodied gecko with short, well-developed limbs. Dorsally, ground color is light dusky tan to rufous beige; dorsally series of paired small, dark spots occur from nape to base of tail and dorsolaterally series of small yellow spots from jowl to hips; bright-orange to pinkish-yellow mark occurs dorsally on sacrum and base of tail. Underside is same color as dorsum but lighter from chin to pelvis and light to bright yellow on tail. Small, equal-sized granular, juxtaposed scales cover head, neck, and trunk dorsally and laterally; scales gradually enlarge ventrally but remain granular. Head has small, smooth, plate-like scales bordering mouth and nares; elsewhere granular scales cover head; chin bears single pair of large postmental scales. Tail is thick, subcylindrical, and gradually tapers to blunt tip; it has granular scales above, grading into small flat, slightly imbricate scales ventrolaterally and slightly larger and smooth scales below; it is not segmented. Fore- and hindfeet have five digits, with first rudimentary on each foot. There are four U-shaped subdigital lamellae on fourth toe of hindfoot. Females lack precloacal and femoral pores; adult males have six to nine precloacal pores (median eight) separated from the femoral pore series on each side; total pore count ranges from 16 to 28 (22).

SIZE: Adult females average larger than males. Females range from 31 to 34 mm SVL (mean 33 mm); males from 28 to 32 mm SVL (30 mm). Tail length is about 65% to 75% of SVL. Hatchling size is 15 to 17 mm SVL.

OCCURRENCE: *H. ganoklonis* is a Palauan endemic and occurs widely throughout the archipelago.

HABITAT CHOICE: This gecko is not a house gecko, although it may live adjacent to houses in landscaping shrubbery through gardens (usually abandoned or poorly attended) to forest.

REPRODUCTION: *H. ganoklonis* bears one to two eggs. Other aspects of its reproduction remain undocumented.

MISCELLANEA: This gecko is a nocturnal forager, searching for its insect and spider prey in shrubs to trees, from trunk to leaf tip.

Hemiphyllodactylus typus   **Indo-Pacific Slender Gecko**
Plate 7, Figure 5

APPEARANCE: Small, slender, elongate-bodied gecko with short, well-developed limbs. Dorsally, ground color is dusky tan to reddish beige; dorsally series of small, dark bars occur from nape to base of tail and laterally series of small whitish spots from jowl to hips; white to beige mark occurs dorsally on sacrum and base tail. Underside is dusky from chin to pelvis and pale orange on tail. Small, equal-sized granular, juxtaposed scales cover head, neck, and trunk dorsally and laterally; scales gradually enlarge ventrally but remain granular. Head has small, smooth, plate-like scales bordering mouth and nares; elsewhere granular scales cover head; chin bears single pair of large postmental scales. Tail is thick, subcylindrical, and gradually tapers to blunt tip; it has granular scales above, grading into small, flat, slightly imbricate scales ventrolaterally and slightly larger and smooth scales below; it is not segmented. Fore- and hindfeet have five digits, with first digit rudimentary on each foot. There are five U-shaped subdigital lamellae on fourth toe of hindfoot. In spite of being females, most adults have precloacal pores (7–12) separated from tiny femoral pores (none to seven).

SIZE: *H. typus* is an all-female species. Adult females range from 29 to 46 mm SVL (mean 38 mm); tail length is about 65% to 75% of SVL. Hatchling size is unknown.

OCCURRENCE: This Slender Gecko occurs spottily throughout the Pacific from Hawaii and French Polynesia westward into coastal Indo-Australia and southward to the Mascarene Islands. This distribution is that of an alien species that has been widely transported from its origin, which is unknown.

HABITAT CHOICE: *H. typus* is a house gecko of unlighted surfaces. It also lives in residential shrubbery, abandoned gardens and similar scrubby area to forest.

REPRODUCTION: All known populations of *H. typus* contain only females. This parthenogenetic gecko lays two eggs in cracks in buildings, leaf axils, and similar protected sites. Incubation is about 60 to 65 days.

MISCELLANEA: The Indo-Pacific Slender Gecko is a nocturnal and arboreal forager, preying on insects from leaf tips to tree trunk.

Lepidodactylus buleli   **Vanuatu Ant-nest Gecko**

Plate 8

APPEARANCE: Small, stout-bodied gecko with well-developed limbs. Dorsally, ground color is light beige with a wash of reddish brown on top of head and posterior trunk; series of five dark-brown, broken bars cross trunk from shoulders to sacrum. Dark stripe extends from snout to eye; broad, golden-tan stripe edged above and below by broad brown stripes extends from eye diagonally downward on the jowls; lips are golden tan. Tail is light brown with dark-brown bands. Venter is mostly white, yellowish on chin and with small brown flecking on abdomen. Small, equal-sized granular, juxtaposed scales cover head, neck, and trunk dorsally and laterally; scales gradually enlarge ventrally but remain granular. Head has small, smooth, plate-like scales bordering mouth and nares; elsewhere granular scales cover head; chin bears series of postmental scales, middle pair slightly larger and grading smaller. Tail is thick, subcylindrical, and gradually tapers to blunt tip; it has granular scales above, grading into small, flat scales ventrolaterally, with slightly larger and smooth scales below; it is not segmented. There are 13 to 14 subdigital lamellae on fourth toe of hindfoot; distal two or three are medially divided. Adult males lack precloacal and femoral pores.

SIZE: This species is known from a single adult male of 38 mm SVL and tail length of 40 mm, and a badly damaged female of about the same size.

OCCURRENCE: Presently, *L. buleli* is known from a single forest site on the west coast of Espiritu Santo, Vanuatu.

HABITAT CHOICE: Several eggs and an adult female were taken from ant-nest structures of epiphytes growing in the forest canopy.

REPRODUCTION: Aside from its oviparity (two eggs in a clutch), reproductive data are not available.

MISCELLANEA: This species has only recently (2008) been described. Only two individuals have been seen by naturalists: an adult female (badly damaged) and an adult male (found as an egg, hatched, and raised to adulthood in captivity).

Lepidodactylus euaensis   **'Eua Forest Gecko**

Plate 8

APPEARANCE: Midsize, stout, somewhat elongate gecko with well-developed limbs. Dorsally, ground color is medium to rufous brown with overlay of darker and lighter marks, creating mottled pattern in

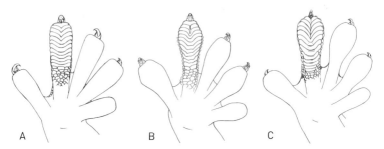

FIGURE 6. Lamellae morphology of the hindfeet of the three morphological groups of *Lepidodactylus.* (A) *L. manni,* Group I member: all digital lamellae are entire, neither strongly cleft or divided; (B) *L. gardineri,* Group II: a few subterminal lamellae are cleft or divided; and (C) *L. lugubris,* Group III: all or many subterminal lamellae are cleft or divided. In addition to the difference in medial division of the lamellae, Group I and II geckos have cylindrical to subcylindrical tails with little or no lateral ornamentation; Group III geckos have dorsoventrally flattened tails, typically with a fringe of projecting scales along the lateral margin.

many individuals; most individuals have lateral series of light spots on neck and trunk and others have dark-brown, transverse bars on back and barely extending onto sides. Almost all individuals bear bright-beige stripe over the sacrum, which is typically bordered anteriorly by dark-brown blotch. Face is usually dark brown with ruby-brown lips. Tail usually has pattern like trunk. Venter is yellow from chin to base of tail and suffused with brown throughout. Small, equal-sized granular, juxtaposed scales cover head, neck, and trunk dorsally and laterally; scales gradually enlarge, flatten, and slightly overlap ventrally. Large, broad head has small, smooth, plate-like scales bordering mouth and nares; elsewhere granular scales cover head; chin bears two to three rows of modest-sized postmental scales, quickly grading into the granular chin scales. Tail is small, subcylindrical, and gradually tapers to blunt tip; it has small, smooth, flat scales above and below, slightly larger below; caudal segments are barely evident. There are 10 to 13 subdigital lamellae on fourth toe of hindfoot, none divided. Adult males have continuous series of 28 to 33 precloacal and femoral pores.

SIZE: Adult females and males appear to be of equal size. Females range from 42 to 50 mm SVL; males from 42 to 47 mm SVL. Tail length is 80% to 90% of SVL. Hatchlings are 21 to 23 mm SVL.

OCCURRENCE: *L. euaensis* is a Tongan endemic, presently known only from the southern islands of 'Eua and 'Ata. Reports from the isolated islands of Niuafo'ou and Late require confirmation.

HABITAT CHOICE: This Tongan gecko is strongly associated with forests with large, older trees and predominantly a closed canopy. Individuals are strongly arboreal and found beneath bark on dead and living trees, usually on upright trees although occasionally living beneath the loose bark of fallen tree.

REPRODUCTION: Females lay clutches of two eggs, always tightly adjacent to one another, and commonly in association with eggs of other females. Females are reproductive October through February; reproductive data for other months are not available. Incubation is about two months.

Lepidodactylus gardineri   **Rotuma Forest Gecko**

Plate 8, Figure 6

APPEARANCE: Midsize, stout-bodied gecko with well-developed limbs. Dorsal and lateral ground color is uniform dark brown with bright-yellow venter except for a bluish-gray chin and anterior throat. This uniform appearance can rapidly lighten to a purplish-gray to light-gray or brown background with broad transverse bars of dark brownish gray from neck onto tail. Venter also lightens but remains yellow. Small, equal-sized granular, juxtaposed scales cover head, neck, and trunk dorsally and laterally; scales gradually enlarge, flatten, and slightly overlap ventrally. Head has small, smooth, plate-like scales bordering mouth and nares; elsewhere granular scales cover head; chin bears series of modest-sized postmental scales of nearly equal size in first row; subsequent rows are progressively smaller grading into granular chin scales. Tail is small, subcylindrical, and gradually tapers to blunt tip; it has small, smooth scales above and below, slightly larger ventrally and slightly overlapping. Tail is not visibly segmented. There are 103 to 118 scale rows around midbody and 12 to 17 subdigital lamellae on fourth toe of hindfoot; distal two to four are medially divided. Adult males have continuous series of 38 to 41 precloacal and femoral pores.

SIZE: Adult females and males are equal in size. Females range from 47 to 50 mm SVL (mean 50 mm); males from 43 to 50 mm SVL (50 mm). Tail length is short, about 60% to 70% of SVL.

OCCURRENCE: *L. gardineri* is a Rotuman endemic.

HABITAT CHOICE: This gecko is a forest inhabitant and is found mainly within termite-hollowed branches of the tree *Acalypha grandis*. Rotuma has large patches of secondary growth forest. The canopy of these forest areas is generally low and closed; the trees are a mixture

locations, reproduction occurs year-round. Incubation is about 40 to 60 days.

MISCELLANEA: Until the mid-1980s, *L. lugubris* was the most common and abundant house gecko throughout its Pacific distribution. With the arrival and spread of the more aggressive *Hemidactylus frenatus*, *L. lugubris* has disappeared from buildings or greatly reduced in number. It continues its association with humans, but now occurs mainly on the shrubs and trees near buildings.

The Mourning Gecko is nocturnal and arboreal. Individuals emerge from their diurnal retreats as it becomes fully dark and eat a full variety of insects, from moths and beetles to ants and termites. They stalk their insect prey to within a few centimeters, and then make a lunging dash to catch the prey.

There are six, perhaps more, clones of *L. lugubris*, which indicates multiple hybridization events for this species. The current data suggest that one parental species was the undescribed Central Pacific Beach Gecko and the other parental species is *L. moestus*; it has been further suggested that hybridization continues. Clone A is most common and widespread of the clones, and distributional data indicate that it occurs at most recorded localities for *L. lugubris*, either alone or with one or more of the other clones. All clones can be recognized by coloration, although only four clones are easily distinguishable (see Plate 9). Each clone has a unique karyotype and genetic signature.

Lepidodactylus manni   **Fiji Forest Gecko**

Plate 8, Figure 6

APPEARANCE: Small, stout-bodied gecko with well-developed limbs and moderate-length tail. Dorsally, ground color ranges from light to medium brown on top of head and posterior trunk; series of five dark-brown, chevron-shaped bars cross trunk from shoulders to sacrum, and black circular, scapular spot is usually present. Dark stripe extends from snout to eye; narrow cream-colored stripe edged above and below by broader brown stripes extends from eye diagonally downward onto jowls. Tail is banded with dark-brown chevrons on lighter-brown background. Venter is uniform yellow from chin to base of tail. Small, equal-sized granular, juxtaposed scales cover head, neck, and trunk dorsally and laterally; scales gradually enlarge ventrally and flatten; these are about four times larger than dorsal scales and lightly overlapping. Head is moderately large, with smooth, plate-like scales bordering mouth and nares; elsewhere granular scales cover head; chin bears pair of moderately large elliptical postmental scales and following scales rapidly grade into granular scales. Tail is thick, subcylindrical, and gradually tapers to blunt tip; it has small smooth scales above and below, slightly larger ventrally and with slight overlap. Tail is not visibly segmented. There are 13 to 14 subdigital lamellae on fourth toe of hindfoot, all lamellae are entire. Males have nine to 14 precloacal pores; occasionally a few femoral pores are present unilaterally.

SIZE: Adult females average larger than males. Females range from 39 to 48 mm SVL; males from 35 to 45 mm SVL. Tail length is subequal to body length, ~90% to 110% SVL. Hatchlings are 16 to 17 mm SVL.

OCCURRENCE: *L. manni* is a Fijian endemic and has been found on Ovalau, Viti Levu, and Kadavu.

HABITAT CHOICE: This gecko is a forest to forest-edge resident. It lives on lichen-covered and bare rock faces and on trees, beneath the bark of dead trees, and in holes and crevices of living trees.

REPRODUCTION: Females lay two eggs. Only one set of eggs has been discovered, and these were within a rock-face crevice.

MISCELLANEA: Our knowledge of the biology of this gecko is nearly nonexistent. As a forest-dweller, its survival is threatened as the original Fijian forests are logged. Presently, there is no evidence that it can survive and reproduce in the pine plantations that have replaced some of the dry grasslands and native dry forest of Viti Levu.

Lepidodactylus moestus   **Micronesia Flat-tailed Gecko**

Plate 9

APPEARANCE: Small, moderately robust lizard with short, well-developed limbs. Head, body, and tail are light to medium brown above with numerous tiny brown spots and occasional rufous-brown spots on neck and trunk. Venter is unicolor white to beige. Small, equal-sized granular, juxtaposed scales cover head, neck, and trunk dorsally and laterally; scales flatten ventrolaterally and are larger and slightly imbricate ventrally. Head has small, smooth, plate-like scales bordering mouth and nares; elsewhere granular scales cover head; chin bears about four rows of small, circular postmental scales. Tail is thick subcylindrical to slightly flattened and gradually tapers to blunt tip; it has granular scales above, grading into small, flat, slightly imbricate scales ventrolaterally; row of elongate scales extends along each ventrolateral edge with larger one projecting outward at posterior edge of nearly every caudal segment; segments are weakly defined. There are 117 to 141 rows of scales around midbody and 10 to 16 subdigital lamellae on fourth toe of hindfoot with terminal three or four lamellae divided. Adult males have 18 to 31 precloacal-femoral pores in continuous series.

SIZE: Adults of this gecko are equal in size. Adult females range from 33 to 39 mm SVL (mean 35 mm); males from 32 mm to 41 mm SVL (36 mm). Tail length is 90% to 100% of SVL. Hatchlings are about 24 to 26 mm SVL.

OCCURRENCE: *L. moestus* occurs in Micronesia from Palau eastward through Pohnpei to Arno and Rongarik Atolls, but not northward into the Marianas.

HABITAT CHOICE: This gecko often co-occurs with *L. lugubris* on buildings and adjacent shrubbery, secondary forest, and seashore scrub. It is strongly arboreal and nocturnal. During the day, it shelters in crevices in buildings and trees and beneath bark.

REPRODUCTION: The reproductive behavior of *L. moestus* remains undocumented. Gravid females commonly have two eggs.

MISCELLANEA: For nearly a century, this gecko was hidden within the nomenclature of the Mourning Gecko. Once the latter species was recognized as an all-female taxon, *L. moestus* and other bisexual populations could be teased apart from their similar-appearing relative *L. lugubris*.

Lepidodactylus oligoporus   **Mortlock Forest Gecko**
Not illustrated

APPEARANCE: Small to midsize, stout-bodied gecko with well-developed limbs and moderate-length tail. Dorsally, ground color ranges from light to medium brown on top of head and posterior trunk; slightly darker-brown stripe runs dorsolaterally from behind eye to hindlimbs; this stripe is bordered above and below by narrow, dark-brown stripes behind eye and on trunk; stripe contains several irregular dark marks. Narrow dark stripe extends from snout to eye. Tail is banded with mottled, dark-brown bars on lighter-brown background. Venter is uniform white from chin to base of tail. Small, equal-sized granular, juxtaposed scales cover head, neck, and trunk dorsally and laterally; Scales gradually enlarge ventrally, flatten and lightly overlap. Head is moderately large with smooth, plate-like scales bordering mouth and nares; elsewhere granular scales cover head; chin bears pair of moderately large elliptical postmental scales and following scales rapidly grade into granular scales. Tail is thick and subcylindrical and gradually tapers to blunt tip; it has small smooth scales above and below, slightly larger ventrally and lightly overlapping. Tail is not visibly segmented. There are 15 to 19 subdigital lamellae on fourth toe of hindfoot, terminal lamella entire, three or four subterminal lamellae divided or strongly cleft. The adult male has 12 precloacal pores.

SIZE: Adult females average larger than males. Females range from 39 to 48 mm SVL; males from 35 to 45 mm SVL. Tail length is subequal to body length, ~90% to 110% SVL. Hatchlings are 16 to 17 mm SVL.

OCCURRENCE: *L. oligoporus* lives only on Namoluk Atoll in the Mortlock Islands.

HABITAT CHOICE: This gecko was found on shrubs in the upper beach zone of a single island.

MISCELLANEA: This gecko was discovered in the mid-2000s and is presently known only from five specimens from a single atoll. Possibly it is more widespread in Mortlock Islands, or at least once was. This species and others confined to the low sandy islands of atolls are on a pathway to extinction because the rising sea level will inundate their islands within a century, or an earlier arriving storm surge will wash the island bare of terrestrial animals.

Lepidodactylus paurolepis　　　　　　　　**Palau Barred Gecko**
Not illustrated

APPEARANCE: Small, stout-bodied gecko with well-developed limbs. Dorsally, ground color is light to medium brown; series of five or six dark-brown, irregularly shaped bars cross trunk from shoulders to sacrum. Broad, dark-brown stripe extends from snout through eye to ear-opening; diffuse dark stripe lies dorsolaterally from eye to mid-temporal area. Tail is light brown above with dark-brown, ill-defined bands. Venter is dusky with suffusion of numerous tiny, dark-brown spots, particularly on throat and sides of abdomen. Small, equal-sized granular, juxtaposed scales cover head, neck, and trunk dorsally and laterally; scales gradually enlarge, flatten, and slightly overlap ventrally. Head has small, smooth, plate-like scales bordering mouth and nares; elsewhere granular scales cover head; chin bears two to three rows of small circular postmental scales, quickly grading into smaller chin granular scales. Tail is subcylindrical and gradually tapers to blunt tip; it has small smooth scales above and below, slightly larger ventrally and lightly overlapping. Tail is lightly segmented. There are 98 to 102 scale rows around midbody and 12 to 14 subdigital lamellae on fourth toe of hindfoot; terminal lamella is entire, three or four subterminal lamellae are divided or strongly cleft. Adult males have continuous series of 31 to 34 precloacal and femoral pores.

SIZE: Adult males range from 38 to 39 mm SVL. Tail length is slightly longer than SVL.

OCCURRENCE: This endemic Palauan gecko is known only from Ngeukeuid Island, although it probably occurs on other nearby islands.

HABITAT CHOICE: The few known specimens of *L. paurolepis* were found in pandanus axils and beneath tree bark in a limestone forest.

REPRODUCTION: Unknown; females probably bear two eggs.

Lepidodactylus tepukapili   Tuvalu Forest Gecko

Not illustrated

APPEARANCE: Midsize, moderately elongate gecko with well-developed limbs. Dorsally, ground color ranges from dark brown to grayish brown on top of head and posterior trunk. Upper surface of neck and trunk has diffuse and lighter-brown mottling. There are no facial markings. Tail is marked with dark-brown chevrons on a lighter-brown background. Venter is uniform yellow from base of neck onto tail; chin and anterior throat is bluish gray. Small, equal-sized granular, juxtaposed scales cover head, neck, and trunk dorsally and laterally. Scales gradually enlarge, flatten, and lightly overlap ventrally. Large, broad head has small, smooth, plate-like scales bordering mouth and nares; elsewhere granular scales cover head; chin bears two to three rows of modest-sized postmental scales, quickly grading into granular chin scales. Tail is small, subcylindrical, and gradually tapers to blunt tip; it has small smooth, flat scales above and below, slightly larger below; caudal segments are barely evident. There are 105 to 118 scale rows around midbody and 12 to 13 subdigital lamellae on fourth toe of hindfoot; terminal lamella is entire, three or four subterminal lamellae divided or strongly cleft. Adult males have continuous series of 38 to 40 precloacal and femoral pores.

SIZE: Adult females and males are probably equal in size. The single known adult female is 41 mm SVL; males range from 43 to 50 mm SVL (45 mm). Tail length is 70% to 80% of SVL.

OCCURRENCE: This gecko has been found only on Funafuti Atoll in Tuvalu.

HABITAT CHOICE: *L. tepukapili* was found beneath bark on standing and downed tree trunks in a coconut woodland dominated by a variety of broad-leafed trees.

MISCELLANEA: This gecko shares a nearly black oral cavity and tongue with its presumed closest relative, *L. gardineri*.

Lepidodactylus vanuatuensis   **Vanuatu Forest Gecko**
Plate 8

APPEARANCE: Small to midsize, stout-bodied gecko with well-developed limbs. Dorsally, ground color is variable and ranges from light gray to medium brown; series of five or six dark-brown bars cross trunk from shoulders to sacrum, lighter-brown on lighter ground color. Facial markings are usually absent from snout to eye; small yellow edging on anterior and posterior borders of eye; contrasting postorbital stripe, usually darker than ground color, runs diagonally from eye to behind jaw. Tail's ground color matches dorsal coloration or slightly lighter, with broad, dark bars. Venter is creamy white. Small, equal-sized granular, juxtaposed scales cover head, neck, and trunk dorsally and laterally; scales gradually enlarge, flatten, and lightly overlap ventrally. The large broad head has small, smooth, plate-like scales bordering mouth and nares; elsewhere granular scales cover head; chin bears two to three rows of small circular postmental scales. Tail is subcylindrical and gradually tapers to blunt tip; it has small smooth scales above and below, slightly larger ventrally and with a slight overlap. Tail is not visibly segmented. There are 99 to 118 scale rows around midbody and 12 to 17 subdigital lamellae on fourth toe of hindfoot, terminal lamella entire, two or three subterminal lamellae divided or strongly cleft. Adult males have continuous series of 38 to 41 precloacal and femoral pores.

SIZE: Adult females are larger than males. Females range from 40 to 47 mm SVL (mean 44 mm); males from 33 to 39 mm SVL (35 mm).

OCCURRENCE: *L. vanuatuensis* occurs widely in Vanuatu.

HABITAT CHOICE: The Vanuatu Forest Gecko lives broadly from the shrubbery adjacent to human dwellings to secondary forest. It also occurs occasionally on the walls of buildings, although seemingly only when shrubbery touches the building.

REPRODUCTION: Females lay two eggs beneath bark, sometimes in communal nests.

MISCELLANEA: Presumably this species eats largely invertebrate prey; however, some individuals have been observed licking the nectar from small flowers.

Lepidodactylus nsp   **Central Pacific Beach Gecko**
Not illustrated

APPEARANCE: Small, moderately robust lizard with short, well-developed limbs. Head, body, and tail are light to medium grayish brown dorsally with slightly darker, broad, transverse bars on the back; top of head has numerous dark spots and narrow, dark-brown postorbital stripe extends from posterodorsal corner of eye to shoulder. Venter is unicolor white to beige. Small, equal-sized granular, juxtaposed scales cover head, neck, and trunk dorsally and laterally; scales flatten ventrolaterally and are larger and slightly imbricate ventrally. Head has small, smooth, plate-like scales bordering mouth and nares; elsewhere granular scales cover head; chin bears about four rows of small, circular postmental scales. Tail is thick, slightly flattened, and gradually tapers to blunt tip; it has granular scales above, grading into small flat, slightly imbricate scales ventrolaterally; row of elongate scales extends along each ventrolateral edge, with larger scales projecting outward at posterior edge of nearly invisible segments. Fourth toe of hindfoot bears nine to 12 subdigital lamellae. Adult males have 11 to 27 precloacal-femoral pores in continuous series.

SIZE: Adult females and males are equal in size. Adult females range from 33 to 41 mm SVL (mean, 38 mm); males from 33 to 42 mm SVL (38 mm). Tail length is ~110% to 125% of SVL.

OCCURRENCE: This Beach Gecko occurs westward from Arno Atoll in the Marshall Islands to Palmayra Atoll and Enderbury into French Polynesia (the Society Islands and Tuamotu), although within this area, the species' actual presence has been confirmed for fewer than 10 islands.

HABITAT CHOICE: This gecko lives solely on atolls, mainly in the scrubby vegetation and tidal wrack on lagoon-side beaches.

REPRODUCTION: Reproductive data are not available. Females probably lay two eggs and are behaviorally similar to *L. lugubris*.

Nactus multicarinatus   **Melanesia Slender-toed Gecko**

Plate 10, Figure 7

APPEARANCE: Midsize, moderately robust lizard with moderately long limbs and narrow digits, without pads on tips. Head, body, and tail are light to medium brown above with amorphous pattern of dark blotches. Venter is unicolor brown from chin onto tail. Small, equal-sized granular, juxtaposed scales cover head, neck, and trunk dorsally and laterally; scales flatten ventrolaterally and are larger and slightly imbricate ventrally; multiple longitudinal rows of keeled, cone-shaped tubercles run on dorsum from nape to base of the tail, 16 to 18 rows at midbody. Pointed, ovate-shaped head has small, smooth, plate-like scales bordering mouth and nares; elsewhere granular scales cover head; chin bears large triangular mental scale bordered laterally by large elliptical postmental scales. Tail is thick subcylindrical and gradually tapers to blunt tip; it has small, slightly keeled imbricate scales above and below, ventral ones barely larger than dorsal ones. No distinct segmentation is evident on tail. All digits are narrow throughout their entire length; there are 20 to 22 subdigital lamellae on fourth toe of hindfoot. Adult males have six to 10 precloacal pores and no femoral ones.

SIZE: Adult females and males are equal in size. Adult females range from 47 to 63 mm SVL; males from 43 to 62 mm SVL. Tail length is 120% to 130% of SVL.

OCCURRENCE: *N. multicarinatus* occurs throughout the northern islands and spottily in the southern islands of Vanuatu. From Vanuatu, its range extends northward into the Solomon Islands.

HABITAT CHOICE: The Melanesia Slender-toed Gecko is largely a forest-floor denizen, resting during the day beneath logs and other forest detritus in secondary to primary forest. At night, it forages in the leaf litter and low on trunks of trees. It also is found in coconut plantations in the husk and frond piles, and within and adjacent to rural dwellings where plant detritus is not regularly burned or removed.

REPRODUCTION: *N. multicarinatus* is a bisexual species. Females lay two eggs, typically in the same locations as their diurnal resting sites, although in "side-tunnels" where they are undisturbed by adult movements. Females appear to lay one clutch of eggs in the dry season and another in the wet season; presumably, each female lays a clutch in each season.

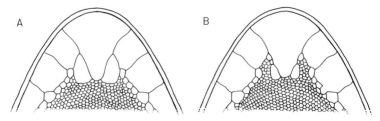

FIGURE 7. Chin scale morphology of Pacific *Nactus*. (A) *N. multicarinatus*. (B) *N. pelagicus*. The large medial mental scale is bordered posteriorly by a large postmental scale on each side in *N. multicarinatus*; postmental scales are absent or small in *N. pelagicus*.

Nactus pelagicus   **Pacific Slender-toed Gecko**

Plate 10, Figure 7

APPEARANCE: Midsize, moderately robust lizard with moderately long limbs and narrow digits, without pads on tips. Head, body, and tail are light to medium brown above with amorphous pattern of dark blotches. Venter is unicolor brown from chin onto tail. Small, equal-sized granular, juxtaposed scales cover head, neck, and trunk dorsally and laterally; scales flatten ventrolaterally and are larger and slightly imbricate ventrally; multiple longitudinal rows of keeled, cone-shaped tubercles run on dorsum from nape to base of tail, 15 to 18 rows at midbody. Pointed, ovate-shaped head has small, smooth, plate-like scales bordering mouth and nares; elsewhere granular scales cover head; chin bears large triangular mental scale bordered laterally by small or no postmental scales. Tail is thick and subcylindrical, and gradually tapers to blunt tip; it has small, slightly keeled imbricate scales above and below, ventral ones barely larger than dorsal ones. No distinct segmentation is evident on tail. All digits are narrow throughout their entire length; there are 21 to 25 subdigital lamellae on fourth toe of hindfoot. As an all-female species, adults lack precloacal and femoral pores.

SIZE: Adults range from 48 to 75 mm SVL. Tail length is 120% to 130% of SVL.

OCCURRENCE: *N. pelagicus* is predominantly a west Pacific resident—that is, from Tonga and Samoa westward. Except for a population on Nassau Island of the Cook Islands, a specimen from Aki Aki (Tuamotu), and a late-19th-century record from Tahiti, this species is absent from central and eastern Polynesia. It occurs throughout Micronesia and in some of the southern islands of Vanuatu, such as Erromango and Tanna.

HABITAT CHOICE: The Pacific Slender-toed Gecko is largely a forest-floor denizen, resting during the day beneath logs and other forest detritus in secondary to primary forest. At night, it forages in the leaf litter and low on the trunks of trees. It is a common resident of the husk and frond piles of coconut plantations.

REPRODUCTION: *N. pelagicus* is an all-female species. Females lay two eggs (occasionally one), typically in the same locations as their diurnal resting sites. Gravid females appear to be present year-round; however, there are no studies of a single insular population to confirm this observation. In Vanuatu, females might lay two clutches each year as *N. multicarinatus* does, one each in dry and wet seasons.

MISCELLANEA: Two presumed bisexual populations have been discovered within the distribution of *N. pelagicus* (Palau and Ngulu Atoll, Yap State). Individuals of both populations are presently indistinguishable from *N. pelagicus* other than the presence of males. Both probably have characteristics that will distinguish them from each other and from *N. pelagicus*; however, no one has yet completed a morphological analysis.

Perochirus ateles   **Micronesia Saw-tailed Gecko**

Plate 6, Figure 8

APPEARANCE: Midsize, robust, slightly flattened lizard with short, heavy limbs. Head, body, and tail are light to medium grayish to greenish brown above with scattering of small, faint, irregular dark markings on neck to base of tail; head largely uniform above and below. Skin tone can darken and lighten and obscure the markings. Venter is grayish white to dusky brown, occasionally with small dark markings. Small, equal-sized granular, juxtaposed scales cover head, neck, and trunk dorsally and laterally; scales gradually flatten ventrolaterally and are larger and slightly imbricate ventrally. The short, pointed head has small, smooth, plate-like scales bordering mouth and nares; elsewhere granular scales cover head; chin bears two rows of postmental scales, midline pair usually largest. Tail is distinctly flattened, yet thick and lanceolate shaped with rounded tip. Its dorsal surface has large, flattened granular scales, ventrolateral edge with small narrow, elongate scales and large, spine-like one projecting outward at the posterolateral edge of each strongly delimited caudal segment. There are 16 to 21 subdigital lamellae (terminal five or seven divided or strongly cleft) on

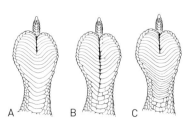

FIGURE 8. Lamellar morphology of hindfoot digital pads in *Perochirus*. (A) *Perochirus ateles*. (B) *Perochirus guentheri*. (C) *Perochirus scutellatus*.

fourth toe of hindfoot. Adult males have o to 5 precloacal pores and no femoral ones.

SIZE: Adult females are presumably smaller than males, but comparative data are not available. Adult males range from about 55 to 88 mm SVL. Tail length is 80 to 100% of SVL. Hatchlings appear to be about 19 to 21 mm SVL.

OCCURRENCE: *P. ateles* occurs in western Micronesia from Tinian, the Mariana Islands, and the Marshall Islands southward through Truk and Pohnpei to Kapingamarangi Atoll.

HABITAT CHOICE: This Saw-tailed Gecko is a forest inhabitant. It is a nocturnal forager and rests during the day beneath bark or within leaf axils.

REPRODUCTION: Limited reproductive data are available; females produce clutches of two to three eggs.

Perochirus guentheri   **Vanuatu Saw-tailed Gecko**

Figures 8, 9.

APPEARANCE: Midsize, robust, slightly flattened lizard with short, heavy limbs. Dorsal ground color of head, body, and tail is probably light to medium grayish brown above with scattering of small, faint, irregular dark markings on neck to base of tail. Skin tone can darken and lighten and obscure the markings. Venter is probably grayish white to dusky brown. Small, equal-sized granular, juxtaposed scales cover head, neck, and trunk dorsally and laterally; scales gradually flatten ventrolaterally and are larger and slightly imbricate ventrally. The short, pointed head has small, smooth, plate-like scales bordering mouth and nares; elsewhere granular scales cover head; chin bears two rows of postmental scales, midline pair usually largest. Tail is distinctly flattened, yet thick and lanceolate shaped with a rounded tip. Its dorsal surface has large, flattened granular scales, ventrolateral edge with small, narrow, elongate scales and large, spine-like one projecting outward at the pos-

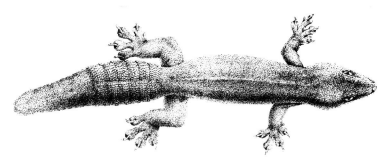

FIGURE 9. Dorsal view of *Perochirus guentheri*. This species has been rarely seen since its discovery in 1860.

terolateral edge of each strongly delimited caudal segment. There are about 14 subdigital lamellae (terminal ones divided or strongly cleft) on fourth toe of hindfoot. Adult males have 10 to 12 precloacal pores and no femoral ones.

SIZE: Adult females are presumably smaller than males, but comparative data are not available. Adult males range from about 61 to 71 mm SVL. Tail length appears to be about 80% of SVL.

OCCURRENCE: *P. guentheri* is or was endemic to southern Vanuatu. Two specimens were collected in 1860 on Erromango and became the type specimens of this taxon, and a third was collected presumably from this island in 1971. A fourth specimen was collected in 1975 on Anatom

HABITAT CHOICE: This Saw-tailed Gecko is presumably a forest inhabitant.

REPRODUCTION: No reproductive data are available, although females probably produce clutches of two eggs.

MISCELLANEA: This species is known only from three individuals with reliable locality data: two collected in 1860 and a third in 1975.

Perochirus scutellatus   **Giant Saw-tailed Gecko**

Figure 8

APPEARANCE: Midsize to large, robust, slightly flattened lizard with short, heavy limbs. Head, body, and tail are light greenish gray to dark gray or brown above with numerous irregular dark areas on head to and onto tail. Skin tone can darken and lighten and obscure the markings. Venter is uniformly grayish white to light greenish yellow. Small, equal-sized granular, juxtaposed scales cover head, neck, and trunk dorsally and laterally; scales gradually flatten ventrolaterally and are larger and slightly imbricate ventrally. The short triangular head with truncate snout has small, smooth, plate-like scales bordering mouth and nares; elsewhere granular scales cover head; chin bears series of modest, circular postmental scales bordering labial scales; subsequent chin scales quickly grade into small granular scales. Tail is distinctly flattened, yet thick and lanceolate shaped with a broad tip. Its dorsal surface is uniformly covered with modest, flattened granular scales, its ventrolateral edge with small narrow, elongate scales and large, spine-like one projecting outward at the posterolateral edge of each strongly delimited caudal segment; ventral scales flat, smooth, and about three to five times larger than dorsal ones. There are 19 to 26 subdigital lamellae (terminal six or seven divided) on fourth toe of hindfoot. Both adult females and males have continuous series of precloacal-femoral pores, although males average more (47–52, mean 52) and larger ones than females (40–53, 47).

SIZE: Adult females are smaller than males. Adult females range from 86 to 114 mm SVL (mean, 98 mm); males from 107 to 132 mm SVL (116 mm). Tail length is 80% to 110% of SVL.

OCCURRENCE: *P. scutellatus* occurs only on Kapingamarangi.Atoll.

HABITAT CHOICE: The Giant Saw-tailed Gecko is a forest inhabitant and preferentially uses larger trees. It is active during the day and uses both trunk and limbs as forage sites.

REPRODUCTION: No reproductive data are available, although females probably produce clutches of two eggs.

MISCELLANEA: Until recently, no species of *Perochirus* were reported from Palau. There are still no confirmed observations of this taxon in the main islands of Palau, although a few individuals of a *Perochirus* have been found on the isolated Southwest Islands and surprisingly have the general appearance of *P. scutellatus*.

Phelsuma grandis   **Madagascar Giant Daygecko**

Plate 10

APPEARANCE: Large, stocky gecko with heavy, well-developed limbs and thick tail tapering to blunt tip. Dorsal ground is various shades and tints of green—regularly bright emerald green—from snout to tip of tail; various patterns of red to orangish-red spots and bars occur on head onto tail; typically present are red stripe extending from snout to eye and cloak of small to medium-sized spots from shoulders onto base of tail; occasionally midline spots fused in middorsal red stripe. Venter is ivory to yellowish white from chin to pubis, often green to light golden yellow on underside of tail. Dorsally, scalation is small, equal-sized granular, juxtaposed scales from head to base of tail; laterally scales gradually enlarge and usually become keeled, ventrolaterally flattening and ventrally smooth, flat, overlapping scales. Head is broad, truncated triangle in dorsal outline with small, smooth, granular scales dorsally, enlarging slightly ventrally with smooth, plate-like scales bordering mouth and nares and larger tuberculate granular on jowls; chin bears three rows of progressively smaller postmental scales, anterior pair largest. Thick tail is rectangular in cross-section and segmented; dorsally it is covered with flatten, lightly keeled, imbricate scales that gradually become larger and smooth laterally and ventrally; large plate scales form longitudinal row midventrally. Digits have paddle-shaped terminal pads with about 16 undivided subdigital lamellae on fourth toe pad of hindfoot. Adult males have continuous series of 38 to 44 precloacal-femoral pores.

SIZE: Males average larger than females. In the Hawaiian Islands, adult females range from 90 to 118 mm SVL (mean, 97 mm); adult males from 85 to 118 mm SVL (105 mm). Tail length is about 95% to 105% of SVL. Hatchlings are reported to range from 62 to 63 mm total length or about 30 to 32 mm SVL.

OCCURRENCE: *P. grandis* is the most recently introduced Daygecko in the Hawaiian Islands, first seen in 1996. This species was intentionally introduced into Manoa Valley, Oʻahu, and has since then spread to other areas of Oʻahu and to Maui and Molokaʻi. It is a native of Madagascar and occurs widely in northern and western regions of that island.

HABITAT CHOICE: In its native habitat, the Giant Daygecko is a diurnal gecko of dry forests with open canopies; there and in the Hawaiian Islands, it has adapted to disturbed habitats that retain some trees and shrubs.

REPRODUCTION: In Madagascar, *P. grandis* is a seasonal breeder with females laying eggs from November to May. In the Hawaiian Islands, males are sexually active year-round. Females also appear to aseasonal there, although egg-laying has so far been confirmed only for December to June. Females usually lay clutches of two eggs in an arboreal site such as beneath bark or in leaf axils, although occasionally on the ground beneath litter. It is likely that females lay multiple clutches each year. In captivity, eggs usually hatch in 48 to 80 days; duration is temperature dependent.

MISCELLANEA: This Daygecko feeds primarily on nectar and invertebrates. It probably preys on smaller geckos as well.

Phelsuma guimbeaui   Orange-spotted Daygecko

Plate 10

APPEARANCE: Midsize, stocky gecko with well-developed limbs and thick tail tapering to blunt tip. Dorsal ground is various shades and tints of green and blue from snout to tip of tail; various patterns of red to orangish-red spots and bars occur on head onto tail; typically present are red U-shaped snout bar, anterior interorbital bar, supraorbital stripe that continues as broken dorsolateral stripe, and numerous bars and spots on dorsum of neck, trunk, and tail. Skin tone can darken and lighten, thus changing the intensity and shade of the markings. Venter is yellowish white on abdomen; chin, throat, and pubis are various shades of yellow. Hatchlings and small juveniles are brown with white speckling, and begin to change to adult coloration when half grown. Dorsally, scalation is small, equal-sized granular, juxtaposed scales from head, to base of tail; laterally scales gradually enlarge and flatten, ventrolaterally flattening; ventrally scales are smooth, flat, and overlapping. Head is truncated triangle in dorsal outline with small, smooth, granular scales dorsally, enlarging slightly ventrally with smooth, plate-like scales bordering mouth and nares; chin bears three to four rows of progressively smaller postmental scales, anterior pair largest. Thick tail is subcylindrical and segmented; dorsally it is covered with flatten, smooth, imbricate scales that gradually become larger and smooth laterally and ventrally; large plate scales form single longitudinal row midventrally. Digits have paddle-shaped terminal pads with 13 to 14 undivided subdigital lamellae on fourth toe pad of hindfoot. Adult males have continuous series of about 26 precloacal-femoral pores.

SIZE: Adult females and males are sexual dimorphic. Adult females range from 57 to 60 mm SVL; males from 64 to 70 mm SVL. Tail length is slightly less or equal to SVL. Hatchlings are 29 to 38 mm total length.

OCCURRENCE: *P. guimbeaui* is a recent arrival in the Hawaiian Islands. It was first observed in Oʻahu in the mid-1980s and presently remains a single geographically limited population. It is a native of Mauritius.

HABITAT CHOICE: In its native habitat, *P. guimbeaui* is predominantly an inhabitant of open scrub forest and scattered coastal trees. In the Hawaiian Islands, it lives in a residential landscape.

REPRODUCTION: The Orange-spotted Daygecko appears to live in small harem-like groups of a male and several females. In Mauritius, this Daygecko has a seasonal reproductive cycle; these data are not available for the Hawaiian population but seasonality probably persists. Females lay two eggs, regularly communally with other females, and typically beneath bark or in deep bark crevices. Incubation usually ranges from 40 to 60 days but can be longer if the eggs are deposited in a cool site.

Phelsuma laticauda   **Golddust Daygecko**

Plate 10

APPEARANCE: Midsize, stocky gecko with heavy, well-developed limbs and thick tail tapering to blunt tip. Dorsal ground is variable, usually shades of green on head, neck, trunk, and limbs; small, golden-yellow spots cover neck and anterior trunk dorsally and laterally; posterior trunk with three elongate orange to red spots followed by smaller red spots. Tail yellowish green with small markings of brown to red; usually brown lateral stripe extends from snout and upper lip to base of tail. Dorsally, head typically has red U-shaped snout bar, anterior interorbital bar, and posterior parietal bar; upper eyelid is blue. Skin tone can darken and lighten, changing the intensity of the color and pattern. Venter is uniform light beige. Dorsally, scalation is small, equal-sized granular, juxtaposed scales from head to base of tail; laterally scales gradually enlarge and usually become keeled, ventrolaterally flattening and ventrally smooth, flat, overlapping scales. Head is broad, truncated triangle in dorsal outline with small, smooth, granular scales dorsally, enlarging slightly ventrally with smooth, plate-like scales bordering mouth and nares, and larger tuberculate granular on jowls; chin bears three rows of progressively smaller postmental scales, anterior pair largest. Thick tail is rectangular in cross-section and segmented; dorsally it is covered with flattened, lightly keeled, imbricate scales that gradually become larger and smooth laterally and ventrally; large plate scales form longitudinal row midventrally. Digits have paddle-shaped terminal pads with 14 to 17 undivided subdigital lamellae on fourth toe pad of hindfoot. Adult males have continuous series of 27 to 29 precloacal-femoral pores.

SIZE: Adult females are smaller than adult males. Females range from 43 to 57 mm SVL (mean 51 mm); males from 40 to 60 mm SVL (55 mm). Tail length is about 110% to 115% of SVL. Hatchlings range from 38 to 42 mm total length.

OCCURRENCE: *P. laticauda* is an alien gecko presently found in the Hawaiian Islands on the islands of Hawai'i Maui, Moloka'i, and O'ahu. Within the last decade it has been introduced into Moorea, French Polynesia. All Hawaiian populations derive from a small number of geckos intentionally introduced by hobbyists in Oʻahu in 1974 and subsequently spread widely. It is a native of the northeast coast of Madagascar and the Comoro Islands.

HABITAT CHOICE: The Golddust Gecko has proved exceptionally adaptable and occurs broadly from city to rural areas with patches of original vegetation, but are found mostly in gardens, orchards, and fields. It often is very abundant in urban and suburban horticultural landscapes.

REPRODUCTION: Females appear to deposit eggs year-round, and probably many females lay multiple clutches; however, males appear to have a cyclic sperm generation cycle. Females lay a clutch of two eggs in above-ground cryptic sites such as leaf axils. Incubation is about 40 to 45 days.

MISCELLANEA: Daygeckos, as the name implies, are active during the day rather than at night. They are arboreal and regularly forage for their insect prey on shrubbery and other low vegetation.

Phyllodactylidae: Leaf-toed Geckos

The leaf-toed geckos comprise a tropical American, African, and Southwest Asian group of geckos with 11 genera and approximately 120 species. Many of the species occur in semiarid and arid areas and are terrestrial or semiarboreal. As their family name and common name indicate, most phyllodactylid geckos have a unique morphology of the digit tip. The leaf shape derives from scale arrangement on the end of each toe. The toes do not have the expanded digital pads of the gekkonid geckos. Rather their digits are slender throughout their entire length, and the terminal pair or two of subdigital lamellae are enlarged, flat scales.

Phyllodactylus barringtonensis   Santa Fé Leaf-toed Gecko

Plate 11

APPEARANCE: Small to midsize, moderately robust lizard with moderate-length limbs ending in digits with leaf-shaped pads. Dorsal ground color is yellowish to brownish gray; there is no middorsal light stripe; irregular dusky blotches and spots occur dorsolaterally and laterally. Head is lighter than trunk and shares diffuse dark blotches above; laterally broad dark stripe extends from snout through eye, above ear-opening to shoulder. Venter is uniformly light yellow. Dorsally and laterally, body surface is granular with juxtaposed scales and smooth,

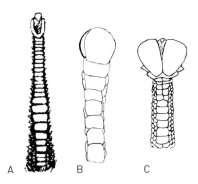

FIGURE 10. Digital lamellae of the fourth toe of Neotropical geckos. (A) *Gonatodes*. (B) *Sphaerodactylus*. (C) *Phyllodactylus*.

imbricate scales anteriorly on limbs and ventrally from midthroat onto tail. Dorsally, trunk lacks rows of trihedral tubercles, although few tubercles occur over the sacral region; limbs also lack tubercles. Head has smooth, plate-like scales bordering mouth and nares, with enlarged granular scales on snout. Tail is cylindrical, tapering gradually to pointed tip with granular scales above, lightly keeled imbricate scales on side and smooth, and subequal sized scales ventrally. Trunk has about 60 to 70 ventral scale rows between fore- and hindlimbs and 10 to 12 subdigital lamellae on fourth toe of hindfoot.

SIZE: Adult average about 41 mm SVL. Tail length is about the same as SVL.

OCCURRENCE: *P. barringtonensis* is an Isla Santa Fé endemic.

HABITAT CHOICE: This gecko is mainly terrestrial, living beneath lava boulders and plant detritus.

REPRODUCTION: Reproductive data are not available, although this gecko probably lays one or two eggs beneath objects on the ground or behind loose bark on standing dead trees.

Phyllodactylus baurii   **Pinta Leaf-toed Gecko**

Plate 11

APPEARANCE: Small, moderately robust lizard with moderate-length limbs ending in digits with leaf-shaped pads. Dorsal ground color is grayish tan; middorsal cream stripe extends from nape onto tail; dusky blotches and spots lie dorsolaterally and laterally. Head is grayish brown with darker-brown smudges above; laterally broad, darkbrown stripe extends from snout through eye, above ear-opening to midneck. Venter is uniformly light yellow to cream. Dorsally and laterally, body surface is granular, juxtaposed scales and smooth, imbricate scales anteriorly on limbs and ventrally from midthroat onto tail. Dorsally, trunk bears 10 to 11 longitudinal rows of trihedral tubercles. Hindlimbs usually lack tubercles on dorsal surface. Head has smooth, plate-like scales bordering mouth and nares and enlarged granular ones on snout. Tail is cylindrical, tapering gradually to pointed tip with granular scales above, lightly keeled imbricate scales on side, becoming smooth beneath; midventrally there is row of enlarged smooth scales. Trunk has about 59 to 79 ventral scale rows between base of neck and vent, 83 to 107 scales around midbody, and 12 to 14 subdigital lamellae on fourth toe of hindfoot.

SIZE: Size data for this gecko are conflicting. One data set gives a maximum SVL of 42 mm, with adult females ranging between 33 to 42 mm SVL; males 34 to 40 mm SVL. Another data set gives means of 44 mm and 47 mm for females and males, with a likely adult range of 40 to 52 mm SVL for both. Tail length is about the same as SVL.

OCCURRENCE: The Pinta Leaf-toed Gecko occurs on Isla Floreana, Isla Pinta, Isla Española, and nearby Gardner.

HABITAT CHOICE: This gecko is mainly terrestrial, living beneath lava boulders and plant detritus in an arid scrub landscape. It has adapted to the human landscape and is abundant around houses.

REPRODUCTION: This gecko lays one or two eggs beneath lava rocks and plant detritus or within rotten tree and cacti stumps. Gravid females were found in March and October.

Phyllodactylus darwini   **Darwin's Leaf-toed Gecko**
Figure 11

APPEARANCE: Midsize, moderately robust lizard with moderate-length limbs ending in digits with leaf-shaped pads. Dorsal ground color is yellowish gray; middorsal light stripe is absent; dusky blotches and spots lie dorsolaterally and laterally. Head is gray with dark mottling; laterally broad, dark stripe extends from snout through eye to above ear-opening. Venter is uniformly yellowish white. Dorsally and laterally, body surface is granular with juxtaposed scales; smooth, imbricate scales anteriorly on limbs and ventrally from midthroat onto tail. Dorsally trunk bears 10 to 12 longitudinal rows of trihedral tubercles. Hindlimbs have tubercles scattered on dorsal surface. Head has smooth, plate-like scales bordering mouth and nares, and enlarged granular scales on snout. Tail is cylindrical and modestly tapered with granular scales above, lightly keeled imbricate scales on side, becoming smooth beneath; midventral row of enlarged, smooth scales. Trunk probably has about 50 ventral scale rows between base of neck and vent, and 13 subdigital lamellae on fourth toe of hindfoot.

SIZE: Males average larger than females. Adult females average ~59 mm SVL; males ~66 mm SVL. Tail length is about the same as SVL.

OCCURRENCE: *P. darwini* lives only on Isla San Cristóbal

HABITAT CHOICE: This gecko is mainly terrestrial, living beneath lava boulders and plant detritus. It also lives in the human landscape.

REPRODUCTION: Reproductive data are not available, although this gecko probably lays one or two eggs beneath objects on the ground or behind loose bark on standing dead trees.

Phyllodactylus galapagensis   Galápagos Leaf-toed Gecko
Plate 11

APPEARANCE: Small to midsize, moderately robust lizard with moderate-length limbs ending in digits with leaf-shaped pads. Dorsal ground color varies from pale cream to medium brown; middorsal light stripe is absent; dusky blotches and spots lie dorsolaterally and laterally. When body darkens, pale lateral spots often appear. Head is uniform gray above; laterally broad dark stripe extends from snout through eye, above ear-opening to shoulder. Venter is uniformly light yellow. Dorsally and laterally, body surface is granular with juxtaposed scales; smooth, imbricate scales anteriorly on limbs and ventrally from midthroat onto tail. Dorsally trunk bears 11 to 12 longitudinal rows of trihedral tubercles in contact with one another fore and aft. Hindlimbs have no tubercles on dorsal surface. Head has smooth, plate-like scales bordering mouth and nares, and enlarged granular scales on snout. Tail is cylindrical, tapering gradually to pointed tip with granular scales above and lightly keeled imbricate scales on side, becoming smooth beneath; midventral row of enlarged smooth scales. Trunk has 65 to 87 ventral scale rows from base of neck to vent, 71 to 107 scales around midbody, and 12 to 14 subdigital lamellae on fourth toe of hindfoot.

SIZE: Adult females and males each average about 42 mm SVL; adult females range from 40 to 45 mm SVL; males from 33 to 46 mm SVL. Tail length is about the same as SVL.

OCCURRENCE: This Leaf-toed Gecko has the broadest distribution in the Galápagos of this group. It is found on Islas Baltra, Barolomé, Cowley,

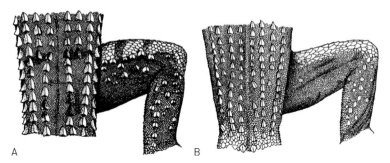

FIGURE 11. Posterior trunk and hindlimb scalation of *Phyllodactylus*. (A) *Phyllodactylus darwini*. (B) *Phyllodactylus reissii;* note absence of tubercles on thigh.

Daphe, Fernandina, Isabela, Pinzón, Plaza Norte, Plaza Sur, Santa Cruz, Santiago, Seymour, Tortuga, and Islote Mares.

HABITAT CHOICE: This gecko is mainly terrestrial, living beneath lava boulders and plant detritus. Some individuals have now become human commensals and live around houses and gardens.

REPRODUCTION: Females lay a single egg, usually in October to November. Eggs are deposited in a variety of sites, such as on the ground beneath stones and plant detritus or beneath bark on dead trees, as well as in holes and crevices in cacti. Because nest sites are often limited, several females may use the same communal nesting site.

Phyllodactylus gilberti   **Wolf Leaf-toed Gecko**

Not illustrated

APPEARANCE: Midsize, moderately robust lizard with moderate-length limbs ending in digits with leaf-shaped pads. Dorsal ground color is grayish brown to mid brown; middorsal light-gray stripe extends from nape onto tail; dusky blotches lie dorsolaterally and laterally. Head is light gray above with irregular, dusky blotches (occasionally absent); laterally, broad dark stripe usually extends from snout through eye, above ear-opening to anterior neck, but stripe is occasionally absent. Venter is uniformly light yellow. Dorsally and laterally, body surface is granular, with juxtaposed scales; smooth, imbricate scales anteriorly on limbs and ventrally from midthroat onto tail. Dorsally trunk bears up to 10, but usually fewer, longitudinal rows of small trihedral tubercles. Hindlimbs have no or few tubercles on dorsal surface.

Head has smooth, plate-like scales bordering mouth and nares, and enlarged granular scales on snout. Tail is cylindrical, tapering gradually to pointed tip with granular scales above and lightly keeled imbricate scales on side, becoming smooth beneath, and midventrally row of distinctly enlarged smooth scales. Trunk has 44 to 48 ventral scale rows between base of neck and vent and 12 to 16 subdigital lamellae on fourth toe of hindfoot.

SIZE: Adults attain lengths up to 56 mm SVL. Tail length is about the same as SVL.

OCCURRENCE: This gecko occurs on Wolf Island, one of the two small islands in the northwest quadrant of the Galápagos Archipelago.

HABITAT CHOICE: Wolf Island is a dry scrub of small trees and cactus. *P. gilberti* lives beneath loose lava rocks. They are common in the vicinity of nesting sea birds, probably owing to an abundance of invertebrate prey.

REPRODUCTION: Reproductive data are not available, although this gecko probably lays one or two eggs beneath objects on the ground or behind loose bark on standing dead trees.

Phyllodactylus leei   San Cristóbal Leaf-toed Gecko
Not illustrated

APPEARANCE: Small to midsize, moderately robust lizard with moderate-length limbs ending in digits with leaf-shaped pads. Dorsal ground color is pinkish tan to light yellowish brown; middorsal light stripe is absent on trunk; trunk is irregularly marked with diffuse dark spots. Head is pinkish tan above; laterally a faded, dark stripe extends from snout through eye, above ear-opening to anterior neck. Venter is uniformly cream to white. Dorsally and laterally, body surface is granular, with juxtaposed scales; smooth, imbricate scales anteriorly on limbs and ventrally from midthroat onto tail. Dorsally trunk and limbs lack tubercles. Head has smooth, plate-like scales bordering mouth and nares, and enlarged granular scales on snout. Tail is lightly tapered cylinder with granular scales above, lightly keeled imbricate scales on side, becoming smooth beneath, and midventrally row of enlarged smooth scales. Trunk has 60 to 70 ventral scale rows between base of neck and vent and 10 to 12 subdigital lamellae on fourth toe of hindfoot.

SIZE: Adult females and males each average 42 mm SVL with a typical range of 38 to 45 mm SVL, and hatchlings are 15 to 16 mm SVL. Tail length is about the same as SVL.

OCCURRENCE: *P. leei* is an Isla San Cristóbal endemic.
HABITAT CHOICE: This gecko lives on the ground beneath plant detritus and rocks, and beneath bark on dead trees in a variety of scrub habitats. It has also successfully adapted to the human-modified landscape.
REPRODUCTION: One to two eggs are deposited beneath lava boulders and have been found in July, October, and November. Multiple clutches are found occasionally in communal nesting sites.

Phyllodactylus reissii   Guayaquil Leaf-toed Gecko

Plate 11, Figure 11

APPEARANCE: Midsize, moderately robust lizard with moderate-length limbs ending in digits with leaf-shaped pads. Dorsal ground color varies from light brown to medium gray; middorsal tan to light-gray stripe extends from nape onto tail; brown to dark-gray blotches and spots lie dorsolaterally and laterally on neck and trunk. Head is tan to gray above with numerous darker, irregularly shaped marks; laterally, broad, dark stripe extends from snout through eye, above ear-opening, and usually to shoulder. Venter is uniformly light yellow to tan. Dorsally and laterally, body surface is granular, with juxtaposed scales; smooth, imbricate scales anteriorly on limbs and ventrally from midthroat onto tail. Dorsally trunk bears 12 to 16 longitudinal rows of closely spaced trihedral tubercles. Tubercles occur on crus (lower leg) and none on thigh. Head has smooth, plate-like scales bordering mouth and nares, and enlarged granular scales on snout. Tail is cylindrical, tapering gradually to pointed tip with granular scales above, lightly keeled imbricate scales on side, becoming smooth beneath, and row of enlarged smooth scales midventrally. Trunk has about 55 ventral scale rows from base of neck to vent and 11 to 17 subdigital lamellae on fourth toe of hindfoot.
SIZE: Adult females average slightly smaller than males. Subadult to adult females on mainland South America range from 40 to 73 mm SVL (mean ~57 mm); males 48 to 75 mm SVL (~59 mm). In Santa Cruz, adults of this alien species show a distinct dimorphism, with males averaging 62 mm SVL and females 56 mm SVL. Hatchlings are about 23 to 24 mm SVL. Tail length usually is about equal to SVL.
OCCURRENCE: *P. reissii* is a mainland South American species from western Ecuador and Peru. It arrived first on Isla Santa Cruz, apparently transported unintentionally to this island, and was reported from

Puerto Ayora in 1975. In early 2011, it was discovered to have dispersed to Isla Isabela. It has also become a resident of the upland town of Bellavista on Santa Cruz.

HABITAT CHOICE: The mainland habitat of *P. reissii* is predominantly arid tropical scrub. There it is mainly terrestrial, living beneath rocks and plant detritus. In the Galápagos Islands, it remains a commensal species confined to towns.

REPRODUCTION: There are no reproductive data for the Galápagos population. On the mainland, females are gravid in November and December, and egg laying likely continues into May. Females lay one or two eggs beneath objects on the ground or behind loose bark on standing dead trees. They are suspected of laying multiple clutches during a single reproductive season.

MISCELLANEA: Presently, *P. reissii* is confined to human settlements, and even though it is larger than its native co-inhabitant, *P. galapagensis*, it has not invaded the latter's natural habitat. It has, however, displaced *P. galapagensis* around and in the towns, so it must be considered a potential threat to the native Galápagos geckos.

Phyllodactylus transversalis   **Malpelo Leaf-toed Gecko**
Plate 11

APPEARANCE: Midsize, moderately robust lizard with moderate-length limbs ending in digits with leaf-shaped pads. Dorsal ground color is grayish brown, and dorsum lacks middorsal light stripe; neck and trunk have broad, transverse, dark-brown bands, two on neck and four on trunk; tail is also dark banded. Head is grayish brown with diffuse brown mottling above, and laterally broad, dark stripe extends from snout through eye, then merging into the mottling on side of head. Venter is uniformly light yellow. Dorsally and laterally, body surface is granular with juxtaposed scales; smooth, imbricate scales anteriorly on limbs and ventrally from midthroat onto tail. Dorsally, trunk bears single pair of longitudinal rows of trihedral tubercles on back. Head has smooth, plate-like scales bordering mouth and nares, and enlarged granular scales on snout. Tail is lightly tapered cylinder with granular scales above, lightly keeled imbricate scales on side, becoming smooth beneath, and midventrally row of enlarged smooth scales. Trunk has 67 to 69 ventral scale rows between base of neck and vent and 13 to 15 subdigital lamellae on fourth toe of hindfoot.

SIZE: Presently data are available for only the holotype, an adult female of 57 mm SVL with a 40 mm regenerated tail.

OCCURRENCE: *P. transversalis* is an Isla Malpelo endemic.

HABITAT CHOICE: Malpelo is a steep sided island, largely grass covered although there are occasional shrubs and patches of herbaceous plants. The gecko probably occurs throughout the island, but it is difficult to observe.

REPRODUCTION: Reproductive data are not available. This gecko likely lays one or two eggs beneath objects on the ground or in rock crevices.

MISCELLANEA: This gecko is nocturnal and seemingly associated mainly with rocky areas and booby nesting sites. It preys on a variety of invertebrate prey associated with rock faces and seabird nesting sites.

Sphaerodactylidae: Minature Geckos

The sphaerodactylid geckos are primarily tropical American species although some occur in Africa and Southwest Asia. There are ten genera and over 200 species, of which nearly 50% are species of *Sphaerodactylus*. The two most speciose genera are *Pristurus* (African) and *Sphaerodactylus* (American); both are terrestrial, have narrow digits, and typically are less than 60 mm SVL.

Gonatodes caudiscutatus   **Shield-headed Gecko**

Plate 11, Figure 10

APPEARANCE: Small, moderately slender gecko with moderate-length limbs with typical lizard narrow and claw-tipped digits. Dorsum is brightly patterned; ground color is light brown overlain by an irregularly edged middorsal light stripe; stripe forks at nape and becomes brighter on the temporal region; neck and trunk have broad, transverse dark-brown bands, interrupted medially by light stripe; tail has alternating bands of dark and light. Often a large white-centered ocellus lies laterally behind the shoulder. Head has a white stripe on each side from snout to above eye and another one from rear of jaw to nape; these light stripes are dark-edged. Venter ground color is cream to beige with few narrow, short, dark lines on chin and anterior throat, distinctly dusky on abdomen owing to numerous tiny brown spots on each scale. Dorsally and laterally, head, body, and tail bear granular, juxtaposed scales; there are smooth, imbricate scales anteriorly on limbs and ventrally from midthroat onto tail. Head has smooth, plate-like scales bordering mouth and nares. Tail is lightly tapered cylinder ending with blunt tip; it bears granular scales above, lightly keeled imbricate scales on side, becoming smooth beneath, with midventral row of enlarged, smooth scales. Trunk has 48 to 52 ventral scale rows between base of neck and vent; there are 18 to 22 subdigital lamellae on fourth toe of hindfoot.

SIZE: In San Cristóbal, females average 38.4 mm SVL; males 40.1 mm SVL.

OCCURRENCE: At the beginning of the 21st century, geckos were the only alien reptiles established in the Galápagos Islands. The Shield-headed Gecko is the oldest alien. It was first reported from Isla San Cristóbal in 1892, and apparently has recently spread to only Baltra if its presence there is verified. It is a native of the coastal lowlands of Ecuador, Colombia, and Peru.

HABITAT CHOICE: *G. caudiscutatus* is an inhabitant of mesic forest and forest edges. It now lives in the moist urban gardens of Puerto Baquerizo Moreno and the wet highland forests.

REPRODUCTION: Reproductive data are not available. All sphaerodactylid geckos are egg-layers, and *G. caudiscutatus* probably has a clutch of two eggs.

MISCELLANEA: Although the Shield-headed Gecko is the oldest alien species, it was considered a new Galápagos species when first captured. Only in 1965 was it recognized as the mainland coastal species *G. caudiscutatus*, hence an alien from mainland South America. Its occurrence in the highland appears to be recent, and it now appears to be well established there. No native lizards occur in this moist forest setting.

Sphaerodactylus pacificus   **Cocos Pygmy Gecko**

Figure 10

APPEARANCE: Midsize, slender-bodied gecko with distinctly pointed head and small, rhomboidal pads on digit tips. Adults are lavender brown with numerous small to large elongate, dark-brown blotches on neck and trunk, and contrasting light-edged dark sacral patch. Venter ground color is white to cream with fine stippling of tiny, dark-brown spots, one on each scale. Dorsally and laterally, head, body, and tail bear granular, juxtaposed scales, some lightly keeled or tuberculate; there are smooth, imbricate scales anteriorly on limbs and ventrally from midthroat onto tail. Head has smooth plate-like scales bordering mouth and nares. Tail is lightly tapered cylinder ending with blunt tip; it bears granular scales above and lightly keeled imbricate scales on side, becoming larger and smooth beneath. There are 13 to 15 subdigital lamellae on fourth toe of hindfoot.

SIZE: Males and females are equal in size. Adults range from 42 to 49 mm SVL, and the thick cylindrical Tail length is about half of total length (~98 to 104% of SVL).

OCCURRENCE: *S. pacificus* is an Isla del Coco endemic.

HABITAT CHOICE: This gecko live within and beneath litter throughout all island habitats from the supralittoral beach wrack to the forest floor in the center of the island. It is assumed to be nocturnal as it feeds on the surface litter at night.

REPRODUCTION: Gravid females bear a single oviducal egg, which is the usual clutch size for *Sphaerodactylus* geckos. Eggs are deposited above ground in tree root and bark crevices, commonly in a communal nest. There are no data on whether the reproductive cycle is year-round or seasonal.

MISCELLANEA: *S. pacificus* is the only Pacific island *Sphaerodactylus*. *Sphaerodactylus* geckos are widespread and diverse in and among the West Indies islands, but less so in mainland Central America and northern South America.

IGUANIA: IGUANAS AND RELATIVES

The Iguania, once considered to be one of the two basal branches of living lizards, is now recognized as one of three branches in a more inclusive and younger phylogenetic branch (Toxicofera) that includes snakes and anguimorphs (monitors, glass lizards, galliwasps, Gila Monster, and others). The species content of the Iguania has not changed with new recognition, however, but it now has a somewhat more recent divergence (although still ancient—it is thought to have diverged during the earliest Cretaceous, approximate 150 million years ago). The Iguania comprises the Agamidae, Chamaeleonidae, and a dozen "iguanid" families, including the true iguanas (Iguanidae), the anoles (Dactyloidae), American fence lizards and relatives (Phrynosomatidae), the Neotropical ground lizards (Tropiduridae), and a variety of other Neotropical families as well as a single Madagascaran family. The Agamidae and Chamaeleonidae are Old World (Eastern Hemisphere) groups and are present on tropical Pacific islands as alien species.

Dactyloidae: Anoles

The family Dactyloidae has recently (in 2011) been resurrected to contain the anoles and only the anoles. Evidence from nuclear and mitochondrial genes reveals the anoles to be an ancient clade distantly related to other Iguania. The anoles, *Anolis*, are totally a New World group of lizard and nearly totally tropical in their distribution. They all share a similar body form: an angular, bony head; a somewhat slender, laterally compressed body; a long tail; and well-developed limbs with unique toe morphology. This body form has proved exceptionally suitable for an arboreal life and has permitted the evolution of more than 360 species. They are particularly abundant and diverse in the West Indies, where multiple species share the same island, and have partitioned the landscape into a half dozen or more life zones—such as the "low tree-trunk zone" or "tree crown zone"—and accordingly have evolved specific lifestyles and body morphologies to fit their preferred habitat.

Anolis agassizii   **Malpelo Anole**

Plate 15

The Malpelo Anole is one of two truly Pacific anoles whose ancestors dispersed naturally over water.

APPEARANCE: Midsize to barely large, slender lizard with moderately long, slender limbs and tail. Dorsal ground color is dark brown to dark blue with numerous small, light-colored spots. Both males and females have reduced dewlaps of same color as adjacent throat; in males, dewlap has bright, median spot; this spot is absent in females. Venter is grayish white to bluish. Head is angular, elongate pentagon in dorsal view and sharply set off from neck. Dorsal surface of neck and trunk has small, abutting granular scales; scales gradually enlarge laterally and ventrally they are lightly keeled and slightly imbricate. Small, irregular-shaped scales cover top and sides of head; supraorbital scales are larger and rugose and form low, paired parenthesis-shaped crest; parietal scale is moderately large and bears parietal eye in its center. Tail is subcylindrical gradually tapering to whip-like tip. Sexually active males have large nuchal crest on neck; there are 48 to 52 (20 to 21 on expanded mid-digit pad) subdigital lamellae on fourth toe of hindfoot.

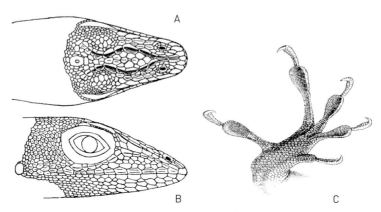

FIGURE 12. Head and hindfoot of anoles. Dorsal (A) and lateral (B) views of the head of *Anolis carolinensis*, and ventral (C) view of the hindfoot of *Anoles equestris*. The heads of anoles typically bear narrow bony ridges and moderate-sized rugose scales. Their digits have a unique pad and lamellae morphology, strikingly different from any native Oceania lizard.

SIZE: Males average larger than females. Adult females range from 71 to 84 mm SVL (mean 77 mm); males from 89 to 114 mm SVL (98 mm). Tail length is ~150 % of SVL.

OCCURRENCE: *A. agassizii* is an Isla de Malpelo endemic and occurs throughout the island, although it is less abundant near the shore and on smooth rock faces.

HABITAT CHOICE: Malpelo is a small, steep-sided island, largely covered in grass although with occasional patches of herbaceous plants and isolated or small clusters of shrubs. There are numerous rock outcrops, and these seem to be the preferred habitat.

REPRODUCTION: Presumably, *A. agassizii* has the usual anoline reproductive behavior of laying a single egg while a second follicle matures in the ovary. Small juveniles have been observed in February and July to September, although another survey found no juveniles in February.

MISCELLANEA: *A. agassizii* is less territorial and more socially tolerant than anoles elsewhere. Individuals aggregate at food and water sources. They are catholic carnivores and prey upon all Malpelo invertebrates with the exception of earthworms. Ants and beetles are the main prey. Evidence suggests that these anoles search for prey as well as being sit-and-wait predators.

Anolis carolinensis   **Green Anole**

Plate 15, Figure 12

APPEARANCE: Midsize, moderately slender lizard with moderately long, strong limbs and tail. Dorsal ground color is uniform bright green that can change to brown or dark gray; in dark phase, dark-brown stripe is sometimes evident extending from beneath eye to shoulder. Male's dewlap is bright pink. Venter is white from chin onto tail. Head is angular, elongate pentagon in dorsal view and sharply set off from neck. Dorsal surface of neck and trunk have small, abutting, lightly keeled granular scales; ventrolaterally and ventrally, scales remain small and are flat, mostly smooth, and slightly imbricate. Small, irregular-shaped scales cover top and sides of head; some supraorbital ones are larger and keeled; parietal scale is small and bears parietal eye. Males have moderate-sized dewlap, which is absent in females. Tail is subcylindrical to lightly compressed and gradually tapers to whip-like tip. Both males and females lack middorsal crest on neck. There are 31 to 38 (20 to 26 on the expanded mid-digit pad) subdigital lamellae on fourth toe of hindfoot.

SIZE: Males average distinctly larger than females. Adult females range from 45 to 61 mm SVL (mean 52 mm SVL); males from 45 to 72 mm SVL (57 mm) (data are from the South Carolina population). Males reach a maximum total length of nearly 130 mm, of which tail is ~60% of total length (~140% to 150 % of SVL). A Guam hatchling was 21 mm SVL.

OCCURRENCE: This southeastern North American species is now widespread on Guam and the Hawaiian Islands (Hawaiʻi, Kauaʻi, Lanai, Maui, Molokaʻi, and Oʻahu). Individuals were first was observed in Oʻahu in 1950, probably as escapees from pet owners because the species was commonly sold as pets in "five-and-dime" variety stores at that time. It rapidly spread to all major islands and became established in each.

HABITAT CHOICE: *A. carolinensis* is an arboreal species of open-forest and forest-edge habitats. It thrives in horticultural landscapes and is seen regularly on bushes and trees around houses.

REPRODUCTION: The anoline ovulation and egg-laying cycle was discovered in a New Orleans population of this species by an embryologist. The anole's courtship occurs in the early spring. The female soon begins to deposit a single egg about every other week throughout the spring and summer. Internally, her two ovaries alternate egg production, with one ovary beginning maturation of an ovum while the other ovary finishes maturation and ovulates the ovum into that side's oviduct. The sequence was initially proposed as very clocklike, with an egg laid every 13 or 14 days. Subsequent research demonstrated that there is less uniformity among individuals, with larger females producing an egg every five to 10 days and smaller (recently matured) individuals requiring 20 to 25 days to produce an egg. When an egg is laid, a female hides it beneath forest floor litter but makes no effort to dig a nest hole. Incubation requires six to eight weeks. A single incubation record from Guam was 30 days.

Anolis equestris   **Knight Anole**

Plate 15, Figure 12

APPEARANCE: Large, moderately robust lizard with moderately long, strong limbs and tail. Dorsal ground color is green overlain with dense black spotting; broad white to light-yellow stripe on upper lip extends from snout to ear-opening; similarly colored stripe extends diagonally upward from shoulder onto anterior trunk. Green ground color can change to dark brown or gray. Dewlap in both females and males is light pink. Venter is white from chin onto tail. Head is angular, elongate pentagon in dorsal view and sharply set off from neck. Dorsal and lateral surface of neck and trunk has smooth, flat scales, each encircled by tiny granular scales. Venter has smooth, flat, slightly imbricate scales that are slightly smaller than lateral scales. Dorsal surface of head in adults has rough appearance with its covering of small to medium-sized, irregular-shaped scales; surface of most scales is rugose; scales of canthal row are enlarged, cone-like, strongly rugose, and form crenulated border, those on rear of head forming casque. There is no enlarged parietal scale. Males have large dewlap, which is present but smaller in females; in both, its outer edge has a border of smooth, round scales. Tail is subcylindrical, gradually tapering to whip-like tip. Males and females have modest middorsal crest on neck. There are 52 to 60 (35 to 40 on the expanded mid-digit pad) subdigital lamellae on fourth toe of hindfoot.

SIZE: Males average slightly larger than females. Adult females range from ~136 to 167 mm SVL, males from ~152 to 179 mm SVL. Tail length is 185% to 205% of SVL. Hatchlings are 80 to 114 mm total length.

OCCURRENCE: The Knight Anole is a western Cuban lizard that was introduced into the Miami area of southern Florida in 1952 and subsequently carried to the Hawaiian Islands. It is now well established on O'ahu.

HABITAT CHOICE: *A. equestris* is strongly arboreal and regularly lives in the upper half of trees. It is a native of open, dry forests and adapted to agricultural, suburban, and urban landscapes with trees that have overlapping branches.

REPRODUCTION: Details of reproduction for both native and alien populations are incomplete. Females lay single, large elliptical eggs. Eggs are laid either in tree holes or in holes dug by females on the ground, usually near the base of a tree. Incubation ranges from 60 to 92 days.

MISCELLANEA: *A. equestris* is an omnivore. It is a predator of insects and small vertebrates; during fruiting seasons, fruit can become a major portion of the lizard's diet. It is both a sit-and-wait predator and a stalker, moving very slowly toward the prey until it is within striking distance then finishes with a rapid rush. It can be aggressive, and defends itself with an open mouth gape and a willingness to bite.

Anolis sagrei   **Brown Anole**

Plate 15

APPEARANCE: Small to midsize, medium-robust lizard with moderately long limbs and tail. Dorsal ground color is brown to grayish brown; adult males are usually uniformly colored, whereas juvenile and females have middorsal dark-brown zigzag pattern bisected by cream stripe from neck to sacrum. Venter is dusky with whitish to light yellow, midline stripe on throat (that is, external edge of dewlap ridge). Male's dewlap is reddish orange. Head is angular, elongate pentagon in dorsal view and merges gradually into neck. Dorsal surface of neck and trunk has small, slightly overlapping unicarinate scales; scales gradually become smaller laterally; ventrally, they are larger, lightly keeled, and slightly imbricate. Small, irregular-shaped scales cover top and sides of head; parietal scale is moderate-sized and bears parietal eye. Males have large dewlap that is present but smaller and slightly extendable in females; in both sexes, its outer edge has border of keeled scales. Tail is laterally compressed with low middorsal crest of upright, flat angular scales, gradually tapering to whip-like tip. Male's middorsal crest is barely more than slight ridge on neck; there are 27 to 32 (nine to 13 on expanded mid-digit pad) subdigital lamellae on fourth toe of hindfoot.

SIZE: Males average larger than females. Adult females range from 34 to 46 mm SVL (mean 43 mm); males from 38 to 70 mm SVL (55 mm). Tail length is ~175% to 220 % of SVL. Hatchlings are 15 to 16 mm SVL.

OCCURRENCE: This anole was sighted first in the early 1980s in a residential area in Oʻahu. It is now widespread throughout Oʻahu and also occurs on Kauaʻi, Maui, and Hawaiʻi. The Brown Anole is native to Cuba and the nearby Bahaman Bank Islands, and supposedly native in Jamaica, the Caymans, and Belize, although these localities might represent early introductions. It arrived in southern Florida, from whence came the Hawaiian stock. The multiple introductions to Florida from different populations of *A. sagrei* have homogenized the genetics of the alien Floridian population.

HABITAT CHOICE: The Brown Anole is cosmopolitan in open habitats from bushy grasslands and horticultural landscapes to scrub forest and forest margins; it seldom lives within closed canopied forest. In the Hawaiian Islands, it remains an urban-suburban resident of horticultural landscapes. It is a trunk-ground anole, but is as likely to sit on a rock as low on a tree trunk to scan for insect prey.

REPRODUCTION: In southern Florida, reproduction of *A. sagrei* is strongly seasonal—April to October—with the greatest reproductive effort made in first third of the reproductive period. *Anolis sagrei* also reproduces seasonally in its native Cuban habitats and in the Hawaiian Islands. In the latter, courtship and egg-laying occurs from January into September. Females lay a single egg at a time; days later they lay another one, generally not in the same place. The eggs require 32 to 45 days to hatch.

MISCELLANEA: The Brown Anole is an insectivore. Both males and females are strongly territorial. In some Hawaiian populations, both males and females develop a rust-colored head in autumn and early winter; the head returns to typical brown by late spring.

Anolis townsendi   **Cocos Anole**

Not illustrated

The Cocos Anole is one of two truly Pacific anoles, whose ancestors dispersed naturally over water.

APPEARANCE: Midsize, slender lizard with moderately long, slender limbs and tail. Dorsal ground color is brown to olive; yellow, black-bordered strip extends on each side from shoulder to groin. Male's dewlap is dull amber; females lack dewlap or bright-colored throat. Venter is yellowish. Head is angular, elongate pentagon in dorsal view and sharply set off from neck. Dorsal surface of neck and trunk has small, abutting granular scales; scales gradually enlarge laterally and ventrally are flat and slightly imbricate. Chest scales are lightly keeled. Small, irregular-shaped scales cover top and sides of head; some supra-orbital scales are larger and keeled; parietal scale is moderately large and bears parietal eye in its center. Males have moderate-sized dewlap, which is only midventral throat ridge in females. Tail is subcylindrical and gradually tapers to whip-like tip. Males have weak middorsal crest on neck. There are 38 to 46 (17 to 20 on the expanded mid-digit pad) subdigital lamellae on fourth toe of hindfoot.

SIZE: Males average slightly larger than females. Adult females range from 41 to 46 mm SVL; males from 40 to 49 mm SVL. Males reach a maximum total length of nearly 130 mm, of which tail length is ~60% of total length (~140% to 150 % of SVL).

OCCURRENCE: *A. townsendi* is an Isla del Coco endemic.

HABITAT CHOICE: Cocos Anoles are abundant and live throughout all levels of the rainforest from the floor and shrubs to treetops (to a height of 10 m). They are active in both shaded and unshaded areas of the forest.

REPRODUCTION: Reproductive behavior of Cocos Anoles is largely undocumented. Males maintain territories, probably overlapping with the territories of one or more female. Females presumably have a typical anoline egg-laying pattern of a single egg deposited alternately throughout an extended reproductive season.

Iguanidae: Iguanas

The Iguanidae family encompasses a diverse group of moderate to very large lizards of eight genera and nearly 40 species. These lizards seemingly are recognized and labeled by everyone as iguanas. Their common name preceded their scientific recognition as group of related lizards, and Linnaeus first used it for the widespread tropical American mainland species. With the exception of the Fijian iguanas, iguanids are strictly a New World group with species living in areas ranging from the deserts of the Southwest (*Dipsosaurus*, the Desert Iguana, and *Sauromalus*, the Chuckwallas) to northern South America. Most iguanas are terrestrial and live in semiarid to arid habitats. Only the widespread *Iguana* and Fijian *Brachylophus* are arboreal. All species are herbivores from hatching to adulthood.

Amblyrhynchus cristatus   **Marine Iguana**
Plate 13

APPEARANCE: Very large, robust lizard with proportionately small head, heavy muscular limbs, and long tail laterally compressed on distal half. Dorsal ground color is very dark brown to black from head to tail; neck, trunk, limbs, and tail with obscure dark-brown blotches; depending on geographic area, trunk and head have various amounts of deep red. Venter is medium brown from midchest onto tail; chin and throat dark brown with some lighter brown mottling. Dorsally and laterally, body from neck onto tail covered with abutting small, spiny, granular scales. Middorsal crest of erect elongate scales that extend from nape to tip of tail. Dorsally, head has small smooth plates on snout and smaller, rugose plates posteriorly. Middorsal crest has 12 to 14 spine-scales on neck, then hiatus, and 80 to 90 flatter spines on trunk. Small, smooth, abutting granular scales cover venter from chin to base of tail, those of trunk slightly larger than anteriorly. Both males and females have femoral pores, smaller in females, 21 to 29 pores in males; male pores are two to four times larger than those of females and there are fewer of them, maximum of 25 pores. There are 27 to 32 subdigital lamellae on fourth toe of hindfoot.

SIZE: Adult males are larger than females; males can be 10 times larger in body mass than females. Body size varies among the different islands depending upon the availability and quality of food. The range of mean SVL for females is 220 to 390 mm; the mean SVL for males is 273 to 548 mm. Tail length is about 140% to 165% of SVL. Hatchling size is ~102 to 116 mm SVL.

OCCURRENCE: The Marine Iguana occurs on all major and most smaller islands in the Galápagos archipelago, although it is patchily distributed along the coasts of all islands.

HABITAT CHOICE: These iguanas are seashore residents of rocky coast. Their preferred coastline includes numerous tidal pools at low tide for intertidal feeding by the smaller individuals. It also includes rock-surface orientation that protects the iguanas from wind, thus allowing for sufficient heating in the mornings and, during the cooler season, to permit daily feeding.

REPRODUCTION: Females lay one to six eggs at the end of 0.5 to 1.0 m burrows; they then backfill the burrows. The nests are dug in sand, generally in beach areas. Multiple females commonly use the same areas because of the limited number of appropriate nesting areas. Nesting occurs from March to April on most islands, and incubation requires three to four months.

MISCELLANEA: *A. subcristatus* is principally herbivorous, feeding on marine alga. Its preferred foods are the various species of red and green algae, which are the most nutritious; brown alga is eaten in the starvation times of El Nino years. On some islands, these iguanas supplement their marine diet with terrestrial succulent plants. They are also known to eat such things as sea lion and iguana feces, and boobies' regurgitates. Juveniles, most females, and subadult males graze in subtidal pools; only the larger females and adult males can harvest algae in the deeper subtidal waters. The limiting factor is water temperature and possession of sufficient mass to slow body-heat loss in the cold Galápagos waters in order to have adequate time to gather enough food without risking hypothermia.

Although seven subspecies have been recognized based on size, color, and head rugosity differences, genetic studies demonstrate a high homogeneity among the islands. Differences in size result from the difference in availability and nutritional quality of the iguana's food at different sites. The genetic homogeneity also indicates that, even though most individuals never leave their "birth" island, a few individuals do, and these inter-island movements allow for gene flow among the various insular populations.

Brachylophus bulabula   **Fiji Banded Iguana**

Plate 12

APPEARANCE: Large, moderately robust lizard with long, sturdy limbs and long, moderately compressed tail. Coloration is sexually dimorphic. Dorsal ground color of both sexes is dark green, usually lighter in females. Males have two broad transverse bands of light green across entire trunk, light-green neck and sacrum bands, and alternating, equal-width bands encircling tail; females usually lack bands and bars, and are nearly uniform in body coloration. Head color is lighter than trunk, particularly on snout; nasal scale is entirely cream to yellow, and iris is red. Dorsally and laterally, body covered with abutting small, spiny, granular scales from neck onto base of tail. Low middorsal crest of small, erect, flat, triangular scales extends from nape onto base of tail. Numerous small, flat, polygonal scales on top and side of head; dorsal ones are heavily pitted; rostral, nasal, and labial scales plate-like and are largest scales of head. Throat has small flexible dewlap. Middorsal crest is sexually dimorphic, with males having longer crest scales; crest scales are continuous series from nape to hips: about 60 flat, triangular scales, declining slightly in size from anterior to posterior. Ventrally, small, abutting granular scales with blunt spines cover chin and throat; thereafter scales of chest and abdomen are imbricate, flat, keeled, and slightly larger. Tail scales are also keeled and larger than ventral trunk ones and larger ventrally than dorsally. Both males and females have femoral pores, which are smaller and fewer in females; total of both sides ranging from 21 to 39, but no precloacal pores. There are 40 to 52 subdigital lamellae on fourth toe of hindfoot.

SIZE: Adult males are larger than females in mass, not in length. Adult females range from 140 to 184 mm SVL; males from 136 to 193 mm SVL. Tail length is 210% to 240% of SVL.

OCCURRENCE: *B. bulabula* has the largest distribution of the three species of *Brachylophus* and lives in the mesic to moist forests of Viti Levu, Vanua Levu, Ovalau, Viwa, and Kadavu and their associated islands.

HABITAT CHOICE: The Fiji Banded Iguana lives in a variety of forest types and in many areas in fragmented forest. It tends to occupy the larger trees and typically feeds in the higher reaches of the trees.

REPRODUCTION: The nesting season is March to April. Females dig horizontal burrows and lay three to four eggs in a row at the bottom. She then fills the burrow, and the eggs incubate for about 35 weeks.

Brachylophus fasciatus   Lau Banded Iguana

Plate 12

APPEARANCE: Large, moderately robust lizard with long, sturdy limbs and long, moderately compressed tail. Coloration is not or barely dimorphic between sexes. Dorsal ground color of both sexes is dark green. Two medium-width transverse bands of light green cross entire trunk; neck has light-green spots, sacrum with narrow, light-greenish bar and narrow, widely separated bands of light green on tail. Head color matches trunk color; nasal scale is orange immediately around nostril and green peripherally, and iris is orange-red. Dorsally and laterally, body is covered with abutting small, spiny, granular scales from neck onto base of tail. Low middorsal crest of small, erect, flat, triangular scales extends from nape onto base of tail. Head has numerous small, flat, polygonal scales on top and side of head; small parietal eye occupies entire parietal scale; dorsal scales are heavily pitted; rostral, nasal, and labial scales are plate-like and largest scales of head. Throat has small flexible dewlap. Middorsal crest is sexually dimorphic, with males having longer crest scales (spines); crest scales are in continuous series from nape to hips, about 60 flat, triangular scales, declining slightly in size from anterior to posterior. Ventrally, small, abutting granular scales with blunt spines cover chin and throat; thereafter scales of chest and abdomen are imbricate, flat, keeled, and slightly larger. Tail scales are also keeled and larger than the ventral trunk ones and larger ventrally than dorsally. Both males and females have femoral pores, smaller and fewer in females; total of both sides ranging from 14 to 27, but no precloacal pores. There are 40 to 50 subdigital lamellae on fourth toe of hindfoot.

SIZE: Adult females range from 135 to 175 mm SVL, males from 145 to 182 mm SVL. Adult males are larger than females in mass but not length.

OCCURRENCE: *B. fasciatus*, the original all-inclusive Fijian iguana, is now recognized as the species comprising the iguana populations occurring on the islands of the Lau group and the introduced population of Tonga.

HABITAT CHOICE: The Lau Banded Iguana is strongly arboreal. The Lau forests are typically dense, low-canopied forests of mixed native and garden-tree species. The Lau Banded Iguana spends its life largely in the trees and rarely descends to the ground except for egg-laying.

REPRODUCTION: Reproductive data are not available for this species.

Brachylophus vitiensis   **Fiji Crested Iguana**

Plate 12

APPEARANCE: Large to very large, moderately robust lizard with long, sturdy limbs and long, moderately compressed tail. Coloration is not sexually dimorphic. Dorsal ground color of both sexes is dark green, typically with blackish overtone in males. Two narrow white transverse bands cross the entire trunk: one is narrow white longitudinal band on neck on each side of middorsal crest, and other is narrow white sacrum band. There are alternating dark-green and white bands encircling tail, with white bands half to third width of green ones. Head color is similar to trunk; nasal scale and edges of adjacent scales are yellowish orange, and iris is gold. Dorsally and laterally, body is covered with abutting small, spiny, granular scales from neck onto base of tail. Low, middorsal crest of small, erect, flat, triangular scales extends from nape onto base of tail. Numerous small, flat, polygonal scales on top and side of head; large parietal eye occupies entire parietal scale; dorsal scales are heavily pitted; rostral, nasal, and labial scales plate-like and largest scales of head. Throat has large, flexible dewlap. Middorsal crest is sexually dimorphic, with males having longer crest scales; crest scales are in continuous series of about 50 spine-like scales from nape to hips; anterior spines are cone-like with recurved tips and become flatter and triangular posteriorly; they also decline in size from anterior to posterior. Ventrally, small, abutting granular scales with blunt spines cover chin and throat; thereafter scales of chest and abdomen are imbricate, flat, keeled, and slightly larger. Tail scales are also keeled and larger than ventral trunk ones and larger ventrally than dorsally. Both males and females have femoral pores, smaller and fewer in females; total of both sides ranges from 12 to 40, but no precloacal pores. There are 40 to 50 subdigital lamellae on fourth toe of hindfoot.

SIZE: Adult males are larger than females in mass although not in length. Adult females range from 180 to 250 mm SVL; males from 195 to 225 mm SVL. Tail length is 210% to 240% of SVL. Hatchlings are about 80 to 85 mm SVL.

OCCURRENCE: The Fiji Crested Iguana occupies the western fringes of Viti Levu and Vanua Levu and their associated near-shore islands. It also occurs in the islands of the Mamanuca and Yasawa groups.

HABITAT CHOICE: *B. vitiensis* is a dry-forest species; these forests are open, patchy, and largely composed of low, scrubby trees.

REPRODUCTION: The nesting season is from March to April. Females dig horizontal burrows and lay three to four eggs in row at the bottom. They then fill the burrows, and the eggs incubate for about 35 weeks.

Conolophus marthae   **Pink Land Iguana**

Plate 13

APPEARANCE: Very large, robust lizard with heavy, muscular limbs and long, subcylindrical tail. Dorsal ground color is pink to pinkish tan from head to base of tail and limbs; series of dark-brown to black transverse bars cross body from anterior trunk onto base of tail; remainder of tail entirely dark. Venter is white with wide black transverse bars from chest to tail base; chin and throat are uniformly pink, with dark mottling on base of throat. Dorsally and laterally, body from neck onto tail covered with abutting small, spiny, granular scales. Very low middorsal crest of small, erect scales extends from nape to tip of tail; this crest has very short blunt spines on adipose hump (largest in males) on neck, then short hiatus, and 80 to 90 low, blunt spines on trunk. Dorsally, small, smooth, circular scales on snout and smaller, nearly flat scales plates posteriorly. Small, smooth, abutting granular scales cover venter from chin to base of tail; those of trunk are slightly larger than anterior granules. Males and females have gular pouch with flat, flabby dewlap. Both males and females (presumably) have femoral pores; total of both sides of one male is 37; there are no precloacal pores.

SIZE: Adult males are larger than females. Maximum reported size is 470 mm SVL. Additional size data have not been reported. Tail length is about 130% of SVL.

OCCURRENCE: C. *marthae* occurs on Volcan Wolf, Isla Isabela.

HABITAT CHOICE: Like their relatives, C. *marthae* lives in an arid landscape of low- to moderate-density scrub.

REPRODUCTION: No data are available.

Conolophus pallidus   Santa Fé Land Iguana

Plate 13

APPEARANCE: Very large, robust lizard with heavy, muscular limbs and long, subcylindrical tail. Dorsal ground color of head to shoulders and forelimbs is beige to light brown, changing to light dusky brown on trunk from shoulder onto tail and hindlimbs. Venter is light to medium brown from midchest onto tail; chin and throat are typically lighter brown. Dorsally and laterally, body from neck onto tail is covered with abutting small, spiny, granular scales. Low middorsal crest of small, erect scales extends from nape to tip of tail. Top of head has small, smooth, elevated, circular scales on snout and smaller, bluntly spinose scales plates posteriorly. Middorsal crest has 12 to 14 spine-scales on neck, then hiatus, and 80 to 90 flatter spines on trunk. Small, smooth, abutting granular scales cover venter from chin to base of tail, those of trunk slightly larger than anterior granules. Males and females have gular pouch with flat, flabby dewlap. Both males and females have femoral pores; total of both sides ranges from 28 to 48 pores; there are no precloacal pores. There are 28 to 32 subdigital lamellae on fourth toe of hindfoot.

SIZE: Adult males are probably than larger than females, although confirming data are not available. Adults range from 325 to 375 mm SVL. Tail length is 125% to 140% of SVL. Hatchlings average about 100 mm SVL.

OCCURRENCE: C. *pallidus* lives only on Isla Santa Fé.

HABITAT CHOICE: The Santa Fé landscape is a dry scrubby one consisting mainly of tree cactus, Palo Santo trees, and two species of low shrubs. Because *C. pallidus* digs refuge burrows for resting and thermoregulation, it requires open areas within the scrub.

REPRODUCTION: *C. pallidus* is in nonbreeding mode from December through mid-August. Courtship begins in late August and continues through mid-October. At that time, females leave their home burrows and territories to join males in the males' territories and share the males' burrows. The females then move to nesting areas that have wide expanses of loose cinder soil in order to dig their nesting burrows, which are approximately 1.5 to 2.0 meters deep. They deposit their egg clutches, averaging about 10 eggs, at the bottom of the burrow; they then fill the burrow. Females guard their burrow site for a few days to a month before returning to their home territories. Peak nesting occurs in late October, but some females lay as late as early December. Incubation requires three to four months, with most hatchlings emerging in mid to late February.

MISCELLANEA: The Santa Fé Iguana is an herbivore like the other Galápagos iguanas. Its primary foods are the leaves and flowers of the shrubs and fruits, flowers, and pads of the opuntia cactus.

Conolophus subcristatus   **Galápagos Land Iguana**

Plate 13

APPEARANCE: Very large, robust lizard with heavy, muscular limbs and long, subcylindrical tail. Dorsal ground color of head to shoulders and forelimbs is bright golden brown; trunk from shoulder onto tail and hindlimbs is dark brown bordered ventrolaterally by lighter brown. Venter is medium brown from midchest onto tail; chin and throat are typically lighter brown. Dorsally and laterally, body from neck onto tail is covered with abutting small, spiny, granular scales. Low middorsal crest of small, erect scales extends from nape to tip of tail, largest scales on neck. Dorsally head has small, smooth, elevated, circular scales on snout and smaller, bluntly spinose scale plates posteriorly. Middorsal crest has 10 to 14 spine-scales on neck, then hiatus, and about 80 to 90 much shorter and laterally flattened spines on trunk. Small, smooth, abutting granular scales cover venter from chin to base of tail, those of trunk slightly larger than anteriorly. Males and females have gular pouch with flat, flabby dewlap. Both sexes have femoral pore series; there are 30 to 36 pores in males, 15 to 20 pores in females;

pore diameter is larger in males than in females. There are about 26 to 30 subdigital lamellae on fourth toe of hindfoot.

SIZE: Adult males are larger than females. Adult females reach approximately 410 mm SVL; males reach about 480 mm SVL; overall adult size ranges from 360 to 480 mm SVL. Tail length is 110% to 125% of SVL. Hatchlings are about 90 to 110 mm SVL.

OCCURRENCE: *C. subcristatus* is the most widespread of the Galápagos land iguanas. It occurs presently on six islands: Baltra (repatriated population), Fernandina, Isabela, Plaza Sur, Seymour South (translocated population), and Santa Cruz; previously, it also occurred on Rabida and Santiago.

HABITAT CHOICE: Like its congeneric relatives, *C. subcristatus* selects areas of low to moderately dense scrub that provides adequate open space to dig its burrow and have access to plants for food. Its diet is the same as that of *C. pallidus*.

REPRODUCTION: Reproductive behavior and activity for *C. subcristatus* is like that of *C. pallidus*. The differences are in the timing of reproduction, and the reproductive period varies from island to island owing to difference in the local climate. The Fernandina *C. subcristatus* population begins its courtship in mid- to late May and has its egg-laying peak in early July. Females lay one clutch of eggs each year, and the clutch varies from seven to 23 eggs. The eggs hatch in October after about 3.5 months of incubation. Females are aggregate nesters, as are all Galápagos iguanas, because of the limited number of nesting sites that have only a few shrubs, trees and rocks, and friable soil for digging burrows. The nesting sites are used repeatedly and are kept largely vegetation free by the annual nesting aggregations, which uproot any seedling scrubs or trees that may have begun to grow after the previous year's nesting.

Iguana iguana   **Green Iguana**

Plate 12

APPEARANCE: Very large, robust lizard with long, muscular limbs and very long, tapering tail. Dorsal ground color is variable; color is commonly grayish green to bright green; other ground colors are gray, brown, rufous, and black. Most individuals have about five dark diffuse transverse bands on trunk between fore- and hindlimbs. Head color is usually uniform and matches body color. Body is covered with amalgam of scale types; dorsal crest from nape onto tail consists of various sized erect, flat, triangular scales (spines) extending from nape onto tail, initial five or six scales small, abruptly largest scales gradually decrease in size to tail; series of short spines on midline of chin and anterior throat; neck and trunk have smooth, flat, granular scales, larger on venter; limbs with slightly overlapping, single-keeled scales; head scales are plate-like and various size; single very large, flat, circular scale (subtympanic) marks the lower margin of the jowl. Middorsal crest has 34 to 73 erect spine-scales (mean, 54) on neck and trunk, and males have more spines than females. Throat skin is loose, folded, and pendulous; diameter of subtympanic scale is nearly as large or as large as the tympanum (~75%–90% of tympanum diameter). Both males and females have precloacal-femoral series, nine to 23 pores (mean 17) in males; male pores are two to four times larger than those of females, and females have fewer pores (maximum 18). There are 29 to 45 subdigital lamellae on fourth toe of hindfoot.

SIZE: Adult males are larger than females. Adult females range from 220 to 350 mm SVL (mean ~325 mm SVL); males from ~220 to 380 mm SVL (~360 mm). Tail length is 200% to 250% of SVL. The preceding SVL lengths are generalized from native populations because there are major differences in adult size among the widespread populations that inhabit different habitats. Presently, average size and range data are not available for the introduced populations.

OCCURRENCE: *I. iguana* was introduced on Oʻahu in the late 1950s or early 1960s. The first report for Maui was in early 2000s, and it is now present in more remote areas. Its dispersal was probably aided by illegal transport by reptile fanciers. These iguanas are established now in Fiji (islands). They were introduced on Qamea by a resort owner in 2000 and quickly became a reproducing population. Subsequently, *I. iguana* has been introduced on other islands, including Taveuni and Koro. The Green Iguana has a broad distribution in the lowlands of Tropical America from Mexico to southeastern Brazil and Paraguay.

HABITAT CHOICE: Although *I. iguana* is predominantly arboreal and lives in dry scrub to evergreen forest, it also lives successfully in arid to semiarid habitats lacking forest. In the latter habitats, it depends on burrows and rock crevices for safe haven.

REPRODUCTION: In the Americas, Green Iguanas have a single reproductive cycle each year. North of the equator, mating occurs December to January, egg laying in February and March, and the hatchlings emerge after about 90 days' incubation in May and June. The female digs a vertical nest burrow and deposits her eggs at the bottom. Clutch size is strongly correlated with body size and physical condition, and ranges from nine to 71 eggs.

MISCELLANEA: The Green Iguana is an herbivore. It feeds primarily on leaves from a large variety of plants. Unlike some other reptilian herbivores that are insectivorous as juveniles, juvenile iguanas are also herbivorous.

Hatchlings are usually uniform bright green. The green is soon replaced by more somber colors of brown, gray, and olive.

Phrynosomatidae: Horned Lizards and Relatives

The phrynosomatid lizards are North and Central American species; only a few species live south of Mexico. They are predominantly small to moderate-sized lizards with only a few species exceeding 200 mm SVL. They are also mainly terrestrial lizards and consist of nine genera and about 140 species. Their greatest diversity is in Mexico and the southwestern United States, thus are mainly semiarid to arid-adapted species.

Urosaurus auriculatus   **Socorro Treelizard**

Plate 14

APPEARANCE: Midsize, robust lizard with moderately robust limbs. Coloration is variable and sexually dimorphic in adults: male dorsal ground color is grayish blue to bright blue (sexually active); females dorsal ground is light brown; dorsally and dorsolaterally both sexes have a series of narrow, dark-brown transverse bars occur from mid neck to sacrum, often interrupted middorsally; limbs and tail similarly banded by dark brown on blue background. Head is vivid dark blue, its scales narrowly edged in dark brown. Ventrally in males, chin and throat are dark grayish blue, neck and anterior chest light brownish blue, remainder of trunk blue with light-brown, diffuse mottling, and pubis and underside of tail light brownish blue; chin, throat, and abdomen of females light brownish blue. Dorsal and lateral scalation of neck and body is mix of small, smooth, granular scales covering most of surface and larger, lightly keeled, slightly imbricate scales forming middorsal row from shoulders to emerge with larger keeled scales of tail; ventrolaterally, scales quickly enlarge and ventral scales are smooth and imbricate with a transverse gular fold of granular scales. Head bears enlarged, plate-like, smooth scales from snout to nape; parietal scale is largest, pentagonal in shape, and with large parietal eye. Tail covered with whorls of large, lightly keeled, imbricate scales. Fourth toe of hindfoot bears nine to 12 subdigital lamellae. Adult males have nine to 13 femoral pores on each side and no precloacal ones.

SIZE: Adult females are smaller than males. Adult females range from ~46 to 58 mm SVL; males from 61 to 65 mm SVL. Tail length is 180% to 200% of SVL.

OCCURRENCE: *U. auriculatus* occurs only on Socorro Island.

HABITAT CHOICE: Socorro Island has two major habitat types: grassland with scattered shrubs and trees, and several different forest types. *U. auriculatus* uses all available habitats and is most abundant in the grassland, where it regularly uses rocks and logs as thermoregulatory and observational sites. In contrast to *U. clarionenis*, it forages mainly on the ground and uses the rocks for short periods of thermoregulatory basking. In the forest, *U. auriculatus* prefers medium-sized trees and rocks for these activities. In all habitats, *U. auriculatus* appears to forage randomly for invertebrate prey, rather than sit and wait, and has developed a loose social hierarchy.

REPRODUCTION: No data are available.

Urosaurus clarionensis   Clarion Treelizard

Not illustrated

APPEARANCE: Midsize, robust lizard with moderately robust limbs. Coloration is variable and sexually dimorphic in adults: male dorsal ground color is bright green with numerous dark markings, females light brown dorsally with fewer dark markings and greenish-tan lateral stripe from ear-opening to hindlimb; basic pattern of neck and trunk is six longitudinal rows of dark spots, spots always widely separated in parasagittal rows, dorsolateral series from near continuous to interrupted stripe, and laterally fewer spots widely separate on trunk only; tail without series of spots; head mottled above and with pair of narrow postorbital stripes. Venter is nearly uniform grayish green with very faint tiny dark spots. Dorsal and lateral scalation of neck and trunk is mix of small, smooth, granular scales covering most of the surface; double longitudinal row of large, keeled, slightly imbricate scales separated by medial row of smaller scales form flat, middorsal crest from midneck to emerge with larger keeled scales of tail; single dorsolateral row of large, keeled scales extends from anterior trunk to hindlimbs; ventrolaterally, scales quickly enlarge and ventral scales are smooth and imbricate. Head bears enlarged, plate-like, smooth scales from snout to nape; parietal scale is largest, pentagonal in shape, with large parietal eye, and patch of large, keeled scales liee in front of ear-opening. Tail covered with whorls of large, lightly keeled, imbricate scales. Fourth toe of hindfoot bears 23 to 28 subdigital lamellae. Adult males have nine to 14 femoral pores on each side and no precloacal ones.

SIZE: Adult males are larger than females. Adult females are 52± mm SVL; males range from 51 to 65 mm SVL. Tail length is ~210% to 250% of SVL.

OCCURRENCE: This species occurs only on Clarion Island.

HABITAT CHOICE: *U. clarionensis* is abundant among the lava outcrops and scrubby areas in the lowlands and rocky ridges. Both males and females spend considerable time on boulders and appear to forage on and in the immediate vicinity of their selected boulder; they are strongly territorial and are sit-and-wait predators. They are absent from any area without boulders. They prey on a variety of invertebrates, especially spiders and insects.

REPRODUCTION: No data are available.

Tropiduridae: Lava Lizards and Relatives

The tropidurid lizards are mainly tropical American residents. They occupy a large geographic area, from the Caribbean islands southward to temperate South America and from tropical rainforest and arid landscapes to mid montane Andes and temperate pampas. They are moderately diverse, with eight genera and more than 120 species. Most tropidurids are small to moderate-sized terrestrial lizards; the majority lives in semiarid to arid landscapes. Only one clade of tropidurid lizards, perhaps the result of a single dispersal event, colonized the Galápagos Islands.

Microlophus albemarlensis   Galápagos Lava Lizard

Plate 14

APPEARANCE: Midsize, robust lizard with moderately robust limbs. Coloration is variable and sexually dimorphic in adults. Dorsal ground color is medium brown; three broad, cream to tan longitudinal stripes extend from neck to base of tail in juveniles and subadult females; dorsolateral stripes disappear in adult males. Dark-brown spots and blotches lie between the stripes and on the sides; large, black to dark-brown throat patch, often red on sides of neck, occurs in breeding males. Adult females develop bright-red head and neck, and less intense red on sides of trunk. Venter is yellowish anteriorly and grayish on belly with dark-brown spots on throat and anterior chest. Body appears rough with moderate-sized, overlapping, keeled scales. Except for ventral scales and enlarged head scales, all scales are strongly keeled, unicarinate; middorsal row forms modest crest from nape onto tail in adult males. Head bears enlarged, smooth scales from snout to nape; single large parietal scale bears parietal eye on its anterior border. Middorsal crest on trunk has 38 to 50 scales, 52 to 66 scales around midbody, and 60 to 74 midventral scale rows.

SIZE: Adult males are larger than females. Males average about 85 mm SVL. Tail length ranges from 125% to 140 % of SVL.

OCCURRENCE: *M. albemarlensis* is the most widely occurring *Microlophus* species in the Galápagos Islands. It lives on the islands of Fernandina, Isabela, Santiago, Rabi, Santa Cruz, Seymour, and Santa Fé and their associated islets; however, see the comment in the Miscellanea section for a different interpretation.

HABITAT CHOICE: This lava lizard, like its relatives, is a terrestrial ecological generalist of semiarid and arid habitats. *M. albemarlensis* lives in dense scrubby habitats to ones that are nearly bare of vegetation.

REPRODUCTION: Courtship and egg-laying begin at the start of the wet-warm season in late December or early January.

MISCELLANEA: Both females and males are strongly territorial and defend their territories against intruders of the same sex. Defense is a push-up display, typically performed more vigorously by males.

Recently, a molecular study of the populations of *M. albemarlensis* shows that those from Islas Santa Cruz, Santa Fé, and Santiago represent three separate genetic lineages. These lineages were given the resurrected species names *M. indefategabilis*, *M. barringtonensis*, and *M. jacobi*, respectively; however, these species were not characterized, thus their resurrection must be substantiated.

Microlophus bivittatus   San Cristóbal Lava Lizard

Plate 14

APPEARANCE: Midsize, robust lizard with moderately robust limbs. Coloration is variable and sexually dimorphic in adults. Dorsal ground color is medium brown; four cream to tan longitudinal stripes extend from neck to base of tail in juveniles and adults, pair each of dorsolateral and lateral stripes. Sides of neck and trunk beneath lateral stripe are bright red in females and barred red and yellow in males. Venter is red tinged yellow anteriorly and grayish white posteriorly in males, and more uniformly tannish in females; breeding males have narrow black stripe across throat. Body appears rough with moderate-sized, overlapping, keeled scales. Except for ventral scales and enlarged head scales, all scales are strongly keeled, unicarinate; middorsal row forms very low crest from nape onto tail in adult males. Head bears enlarged, smooth scales from snout to nape; single large parietal scale bears parietal eye on its anterior border. Middorsal crest on trunk has 41 to 51 scales, 53 to 64 scales around midbody, and 62 to 77 midventral scale rows.

SIZE: Adult males are slightly larger than females. Males average about 70 mm SVL. Tail length ranges from 125% to 140 % of SVL.

OCCURRENCE: *M. bivittatus* is an Isla San Cristóbal endemic.

HABITAT CHOICE: This lizard is a terrestrial ecological generalist of semi-arid and arid habitats.

REPRODUCTION: Courtship and egg-laying begin at the start of the wet-warm season in late December or early January.

MISCELLANEA: Both females and males are strongly territorial and defend their territories against intruders of the same sex. Defense is a push-up display, typically performed more vigorously by males.

Microlophus delanonis   **Española Lava Lizard**

Plate 14

APPEARANCE: Midsize to large, robust lizard with moderately robust limbs. Coloration is variable and sexually dimorphic in adults. Dorsal ground color is olive brown in males and females; adult males have six to eight longitudinal rows of dark-brown spots extending from neck toward tail. In males, side of head and neck is brown, side of trunk is reddish. Spots are less intense in females, and are orangish red on head, neck, and ventrolaterally on trunk. Venter is light yellowish brown; both sexes have a large black to dark-brown throat patch and black antehumeral spots; males have a spotted chin. Body appears rough with moderate-sized, overlapping, keeled scales. Except for ventral scales and enlarged head scales, all scales are strongly keeled, unicarinate; middorsal row forms modest crest from nape onto tail in adult males and lower crest in females. Head bears enlarged, smooth scales from snout to nape; single large parietal scale bears parietal eye on its anterior border. Middorsal crest on trunk has 50 to 61 scales, 64 to 77 scales around midbody, and 71 to 86 midventral scale rows.

SIZE: Males are larger than females. Adult females range from 83 to 111 mm SVL (mean 95 mm SVL); males reach 138 mm SVL. Tail length ranges from 125% to 150% of SVL. This is the largest species of lava lizard.

OCCURRENCE: *M. delanonis* lives only on Isla Española and adjacent islets.

HABITAT CHOICE: This lizard is a terrestrial ecological generalist. Española is an arid island with a uniform cover of low scrub and scattered patches of grass and bare soil.

REPRODUCTION: Nesting begins at the start of the wet-warm season in late December or early January and continues for several months, stopping with the end of the rains. The females migrate from their inland territories to the coastal beaches, moving as much as 200 to 300 meters. The females dig a nest in the beach sand with their fore- and hindfeet. Nest chambers or tunnels are 10 to 45 cm long. At the bottom of the chamber, the female deposits three to seven eggs, fills the chamber, and returns to her territory. She may repeat this nesting behavior every three to four weeks if prey quantity and quality provide sufficient nutrition to complete another vitellogenic cycle.

MISCELLANEA: Both females and males are strongly territorial and defend their territories against intruders of the same sex. Defense is a push-up display, typically performed more vigorously by males.

M. delanonis is an omnivore, eating flowers and leaves even when its insect prey is abundant.

Microlophus duncanensis   **Pinzón Lava Lizard**
Not illustrated.

APPEARANCE: Midsize, robust lizard with moderately robust limbs. Dorsal ground color is medium brown with heavy scattering of small, dark spots on neck to midtrunk, replaced posteriorly by smaller white spots in males; females have fewer spots and are bright red on mandible and side of head, continuing ventrolaterally onto tail. Venter is reddish tan with numerous dark-brown spots on throat and anterior chest and extending ventrolaterally onto trunk; breeding males have large, black throat patch. Body appears rough with moderate-sized, overlapping, keeled scales. Except for ventral scales and enlarged head scales, all scales are strongly keeled, unicarinate; middorsal row forms modest crest from nape onto tail in adult males, lower in females. Head bears enlarged, smooth scales from snout to nape; single large parietal scale bears parietal eye on its anterior border. Coloration is variable and sexually dimorphic in adults. Middorsal crest on trunk has 44 to 59 scales, 76 to 92 scales around midbody, and 78 to 92 midventral scale rows.

SIZE: Adult males are slightly larger than females. This species is the smallest lava lizard; males average about 65 mm SVL. Tail length ranges from 125% to 140% of SVL.

OCCURRENCE: This lava lizard occurs only on Isla Pinzón and adjacent islets.

HABITAT CHOICE: This lizard is a terrestrial ecological generalist of semi-arid and arid habitats.

REPRODUCTION: Courtship and egg-laying begin at the start of the wet-warm season in late December or early January.

MISCELLANEA: Both females and males are strongly territorial and defend their territories against intruders of the same sex. Defense is a push-up display, typically performed more vigorously by males.

Microlophus grayii   **Floreana Lava Lizard**

Plate 14

APPEARANCE: Midsize, robust lizard with moderately robust limbs. Coloration is similar in females and males. Dorsal ground color is medium brown, unmarked on head and marked with dark-brown to black spots and crossbars on posterior neck, trunk, limbs, and tail. Venter is greenish with a grayish throat and anterior chest. Body appears rough with moderate-sized, overlapping, keeled scales. Except for ventral scales and enlarged head scales, all scales are strongly keeled, unicarinate; middorsal row forms a modest crest from nape onto tail in adult males. Head bears enlarged, smooth scales from snout to nape; single large parietal scale has parietal eye on its anterior border. Middorsal crest on trunk has 36 to 48 scales, 58 to 74 scales around midbody, and 66 to 77 midventral scale rows.

SIZE: This lava lizard is one of the smallest in the Galápagos, averaging about 65 mm SVL. Size dimorphism between males and females is slight. Tail length ranges from 125% to 140% of SVL.

OCCURRENCE: This species occurs only on Isla Floreana and its neighboring islets.

HABITAT CHOICE: This lizard is a terrestrial ecological generalist that lives among broken lava rocks and cactus scrub.

REPRODUCTION: Courtship and egg-laying begin at the start of the wet-warm season in late December or early January.

MISCELLANEA: Both females and males are strongly territorial and defend their territories against intruders of the same sex. Defense is a push-up display, typically performed more vigorously by males.

Microlophus habelii   Marchena Lava Lizard

Not illustrated

APPEARANCE: Midsize to large, robust lizard with moderately robust limbs. Coloration is variable and sexually dimorphic in adults. Dorsal ground color is dark brown with numerous small, gray spots in males and dusky-green spots in females. Both sexes have dark-red sides of neck and trunk. Venter is also red from chin to anterior chest in both sexes and light brown posteriorly. Body appears rough with moderate-sized, overlapping, keeled scales. Except for ventral scales and enlarged head scales, all scales are strongly keeled, unicarinate; middorsal row forms a strong crest from nape onto tail in adult males. Head bears enlarged, smooth scales from snout to nape; single large parietal scale bears parietal eye on its anterior border. Middorsal crest on trunk has 54 to 59 scales, 71 to 79 scales around midbody, and 76 to 81 midventral scale rows.

SIZE: Adult males are slightly larger than females. Males average about 110 mm SVL. Tail length ranges from 125% to 140% of SVL.

OCCURRENCE: *M. habelii* is endemic to Isla Marchena.

HABITAT CHOICE: This lizard is a terrestrial ecological generalist of semi-arid and arid habitats.

REPRODUCTION: Courtship and egg-laying begin at the start of the wet-warm season in late December or early January.

MISCELLANEA: Both females and males are strongly territorial and defend their territories against intruders of the same sex. Defense is a push-up display, typically performed more vigorously by males.

Microlophus pacificus   Pinta Lava Lizard

Not illustrated

APPEARANCE: Midsize, robust lizard with moderately robust limbs. Coloration is variable and strongly dimorphic in adults. Dorsal ground color is grayish brown in males, and neck and anterior trunk bear transverse series of black bars, becoming less defined posteriorly; adult females are brick red from head to midtrunk. Both sexes have light middorsal stripe from nape onto tail. Venter is olive gray; males have pair of dark brown spots on anterior chest. Body appears rough with moderate-sized, overlapping, keeled scales. Except for ventral scales and enlarged head scales, all scales are strongly keeled, unicarinate; middorsal row forms modest crest from nape onto tail in adult males. Head bears enlarged, smooth scales from snout to nape; single large parietal scale bears parietal eye on its anterior border. Middorsal crest on trunk has 43 to 55 scales, 66 to 75 scales around midbody, and 68 to 80 midventral scale rows.

SIZE: Adult males are slightly larger than females. Males average about 100 mm SVL. Tail length ranges from 125% to 140% of SVL.

OCCURRENCE: *M. pacificus* is endemic to Isla Pinta.

HABITAT CHOICE: This lizard is a terrestrial ecological generalist of semi-arid and arid habitats.

REPRODUCTION: Courtship and egg-laying begin at the start of the wet-warm season in late December or early January.

MISCELLANEA: Both females and males are strongly territorial and defend their territories against intruders of the same sex. Defense is a push-up display, typically performed more vigorously by males.

SCINCOMORPHA: SKINKS AND RELATIVES

The scincomorphan lizards are a diverse group comprising the Cordylidae, Gerrhosauridae, Xantusiidae, and seven families recently (in 2012) partitioned from the former Scincidae. The members of the first two families, the Girdled and Plated Lizards, are sub-Saharan lizards, predominantly grassland and savanna residents. Almost all of their species are heavily armored with moderately large, keeled body scales underlain by bony plates (osteoderms). In contrast, the Xantusiidae or Night Lizards have small, granular scales on their backs and sides and lack osteoderms. Xantusiids have a small and scattered distribution from the southwestern United States to northern Panama. The seven families of skinks are the Acontidae, the Egerniidae, the Eugongylidae, the Lygosomidae, the Mabuyidae, the Scincidae, and the Sphenomorphidae. These skink families contain a variety of body forms and life styles. Most frequently they have smaller and smoother scales than the girdled lizards; however, their body scales also are underlain by osteoderms. Skinks occur worldwide, although only the Mabuyidae occurs in the tropical and subtropical areas of all continents, and only peripherally in Oceania. Within the Pacific, the eugongylid skinks are dominant and widespread.

The newly recognized families of skinks derive from long-recognized subfamilies or groups within the former Scincidae. The generic content of these groups and subfamilies has changed over time, and it is likely that the generic content of each family will continue to be refined by additional phylogenetic studies. Such changes in classification are an ongoing process in science as we gather new evidence and perform new analyses. The preceding multiple families of the Iguania were proposed more than two decades ago and have gained widespread acceptance only in the past five years or so. The new familial classification of skinks will similarly be slow to gain widespread adoption.

Eugongylidae: Pacific Skinks

Most Pacific skinks belong to this family. Overall, the distribution of the eugongylids is Australopapuan and the Central and West Pacific, with a few species occurring northward through Islands Asia. They range from small taxa of 30 to 40 mm SVL adults to larger species of nearly 200 mm SVL, although the majority of taxa reach lengths of less than 100 mm. They are arboreal species to subterranean ones, bright to somber in color, and abundant to rare. Reproduction includes both egg-laying and live-bearing; in the latter mode, the species frequently exhibits placental development and nutrient-gas exchange between fetus and mother.

Carlia ailanpalai   **Admiralty Brown Skink**

Plate 22

C. ailanpalai and *C. tutela* differ from all other Pacific skinks by their four-digit forefeet.

APPEARANCE: Midsize, medium-bodied lizard with moderately long limbs and long, tapering tail. Its head is ovate in dorsal outline and distinct from neck. Dorsal ground color is uniform shiny coppery brown to orangish brown, occasionally dark brown. Head matches body coloration with lips lighter brown. Venter from chin onto trunk is uniform cream to light tan. Neck, trunk, and tail have large smooth, imbricate scales; scales are subequal in size around body. Scales on head are smooth, enlarged, abutting plates; supranasal scales are absent and interparietal scale is small, triangular, with parietal eye and

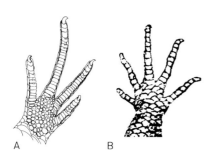

FIGURE 13. Forefoot morphology of Pacific skinks. (A) Four-fingered forefoot of *Carlia fusca*. (B) Five-fingered forefoot of *Eutropis multicarinata*, a typical condition of most Pacific skinks.

rarely absent; there are seven supralabials, fifth is largest and beneath eye. First pair of middorsal body scales behind parietals is enlarged as nuchal plates. Middorsal scales are in 45 to 51 rows from nape to base of tail, 30 to 36 scale rows encircle body in middle of trunk, and fourth toe (hindfoot) bears 26 to 34 lamellae on its underside.

SIZE: Males average slightly larger than females and attain a maximum total length of ~108 mm, of which tail length is about 50% to 55% of total length. Adult females range from 49 to 58 mm SVL (mean 53 mm); males from 46 to 59 mm SVL (54 mm). Hatchlings range from 22 to 25 mm SVL.

OCCURRENCE: This alien species occurs on Guam, Rota, Tinian, and Saipan in the Marianas.

Reports from Yap and Kosrae have not confirmed that the Brown Skink is established there. This species occurs naturally in the Admiralty Islands, off the north coast of Papua New Guinea. Presumably, it was carried to Guam or Saipan with the transfer of war materials after World War II. Its presence on other islands of the Marianas and Micronesia probably derives from this initial unintentional introduction.

HABITAT CHOICE: This skink is strongly terrestrial, uncommonly climbing low on tree trunks or shrubs. It occurs most abundantly in open habitats, grassland, and gardens to open scrub. Although it can be abundant along the edge of forests, it rarely penetrates far into a well-canopied forest.

REPRODUCTION: Gravid females bear two eggs, one in each oviduct. Eggs are deposited shallowly in the soil beneath leaf litter or almost any item large enough to retain a moist environment. Communal nests have been found with various associations—*C. ailanpalai* only or in association with *Gehyra oceanica*, *Anolis carolinensis*, and/or *Emoia caeruleocauda*. There are no data on year-round reproduction, although it is likely. Incubation is 34 to 35 days.

MISCELLANEA: The Admiralty Brown Skink is an active surface forager and quickly investigates any disturbance on the forest floor that might be prey, hence giving rise to the colloquial name "curious skink" in Guam.

Even though the Admiralty Brown Skink is a recent arrival in the Marianas, this skink now dominates the terrestrial lizard fauna of Guam and nearby islands. It appears to have out-competed the previously dominate *Emoia* skinks, perhaps by its more aggressive foraging behavior.

Carlia tutela   **Halmahera Brown Skink**
Not illustrated

APPEARANCE: Midsize, medium-bodied lizard with moderately long limbs and long, tapering tail. Its head is ovate in dorsal outline and distinct from neck. Dorsal ground color is uniform medium to dark brown in males; some adult females retain juvenile light lateral stripe. Head matches body coloration, with lips lighter brown. Venter from chin onto trunk is uniform cream to light tan. Neck, trunk, and tail have large smooth, imbricate scales; scales are subequal in size around body. Scales on head are smooth, enlarged, abutting plates; supranasal scales are absent, and interparietal scale is small, triangular, with parietal eye and rarely absent; there are seven supralabials, fifth one is largest and beneath eye. First pair of middorsal body scales behind parietals is enlarged as nuchal plates. Middorsal scales are usually in 44 to 51 rows from nape to base of tail, 30 to 37 scale rows encircle body in middle of trunk, and fourth toe (hindfoot) bears 23 to 32 lamellae on its underside.
SIZE: Adult males and females are equal in size. Adult females range from 39 to 51 mm SVL (mean, 46 mm SVL); males from 41 to 54 mm SVL (46 mm). Tail length is subequal to SVL.
OCCURRENCE: *C. tutela* is a resident of Halmahera and nearby islands. When this taxon was recognized as a new species, it was thought to be the species that was introduced in Palau during or immediately subsequent to World War II; however, a recent molecular study indicates that the Palau population is genetically identical to individuals of *C. ailanpalai* from the Admiralty and Mariana Islands.
HABITAT CHOICE: This skink is strongly terrestrial, uncommonly climbing low on tree trunks or shrubs. It occurs most abundantly in open habitats, grassland, and gardens to open scrub. Although it can be abundant along the edge of forests, it rarely penetrates far into a well-canopied forest.

REPRODUCTION: Gravid females bear two eggs, one in each oviduct. Eggs are deposited shallowly in the soil beneath leaf litter. There are no data on a year-round or seasonal reproductive cycle.

Cryptoblepharus eximius   **Fiji Snake-eyed Skink**

Plate 16, Figure 14

APPEARANCE: Small, slightly elongate-bodied lizard with short, well-developed limbs and long, subcylindrical tapering tail. Its head is ovate in dorsal outline and barely distinct from neck. Eyes lack moveable eyelids and, as with snakes, each eye is protected by clear scale. Dorsal ground color is variable although consistent within local populations, from golden tan or light gray to black; this ground color forms a broad dorsal stripe from snout onto tail, generally copper colored on head and becoming less iridescent on neck and posteriorly; dorsal stripe narrowly edged in black, silvery-beige dorsolateral stripe from eye to midbody, broad, dark-brown to black lateral stripe from loreal area onto tail. Venter is usually grayish olive with irregular black flecking. Neck, trunk, and tail have large smooth, imbricate scales above and below; middorsal pairs are enlarged, about one and a half times larger than lateral and ventral scales. Scales on head are smooth, enlarged, abutting plates; supranasal and interparietal scales are absent; there are typically seven supralabials, fifth one is largest and beneath eye. First one or two pairs of middorsal body scales behind parietals are greatly enlarged as nuchal plates. Middorsal scales are usually in 46 to 61 rows (usually < 50) from nape to base of tail, 21 to 27 scale rows encircle body in middle of trunk, and fourth toe (hindfoot) bears 16 to 26 lamellae on its underside.

SIZE: Adult females and males are equal in size. Adult females range from 33 to 40 mm SVL (mean 36 mm); males from 33 to 38 mm SVL (35 mm). Tail length is 110% to 125% of SVL. Hatchlings are about 20 mm SVL.

OCCURRENCE: *C. eximius* occurs widely but spottily throughout the Fijian islands, although not in Rotuma.

HABITAT CHOICE: Fiji Snake-eyed Skinks are predominantly seaside inhabitants where creviced rocky surfaces or large wood jams are above the high-tide line. They also can be found inland in areas of large boulders, such as spottily in the Nausori Highlands or on isolated concrete gun bunkers kilometers from the shoreline.

REPRODUCTION: The reproductive biology of this species has not been studied, although gravid females carry one (predominantly) to two

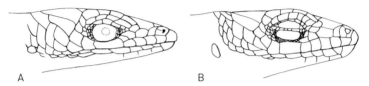

FIGURE 14. Eyelid morphology.of Pacific skinks. (A) *Cryptoblepharus eximius.* (B) *Emoia adspersa.* In the snake-eyed skinks, *Cryptoblepharus,* the lower eyelid is immovably fused at the top of the orbit, and a single scale—the spectacle—covers the eye and is transparent. In many Pacific skinks, the lower eyelid is large and the most movable lid; when the lid closes, the lizard can still see because of a transparent palpebral disc in the middle of the lower lid.

eggs. Nesting sites are uncertain; single communal nest of presumed *C. eximius* eggs was found beneath a rock slab. Reproduction may be seasonal; gravid females were found in May, August, and September.

MISCELLANEA: Daily activity and prey is probably the same as the Oceania Snake-eyed Skink.

Cryptoblepharus nigropunctatus   **Bonin Snake-eyed Skink**

Not illustrated

APPEARANCE: Midsize, slightly elongate-bodied lizard with short, well-developed limbs and long, subcylindrical tapering tail. Its head is ovate in dorsal outline and barely distinct from neck. Eyes lack moveable eyelids, and, as with snakes, each eye is protected by clear scale. This somber-colored skink is seemingly dusky medium brown to gray when viewed from a distance. Closer, a broad brown, dorsal stripe containing dense, dark-brown speckling extends from snout onto tail; this stripe is edged laterally by broken whitish dorsolateral stripe (series of spots) from eye to midbody; and broad, dark-brown lateral stripe extends from loreal area onto tail. Venter is usually brown; often chin and neck are darker than abdomen. Neck, trunk, and tail have large, smooth, imbricate scales above and below; middorsal pairs are enlarged, about one and a half times larger than lateral and ventral scales. Scales on head are smooth, enlarged, abutting plates; supranasal and interparietal scales are absent; there are typically seven supralabials, fifth one is largest and beneath eye. First and second pairs of middorsal body scales behind parietals are greatly enlarged as nuchal plates. Middorsal scales are usually in 47 to 56 rows from nape to base of tail, 22 to 26 scale rows encircle body in middle of trunk, and fourth toe (hindfoot) bears 19 to 27 lamellae on its underside.

SIZE: Adult females and males are equal in size. Adult females range from 37 to 55 mm SVL (mean, 49 mm); males from 46 to 55 mm SVL (50 mm). Tail length is 140% to 160% of SVL.

OCCURRENCE: This skink is endemic to the Bonin Islands, specifically living on Chichijama and Hahajima (Ogasawara Islands) and Minami-Ioujima (Iou Islands).

HABITAT CHOICE: *C. nigropunctatus* resides primarily in grasslands and forest-edge habitats, and uncommonly along the coast. It is a terrestrial forager.

REPRODUCTION: The reproductive biology of this species has not been studied, although gravid females carry one to two eggs.

MISCELLANEA: Daily activity and prey type is similar to that of the Oceania Snake-eyed Skink. However, being an inland species, insects rather than amphipods and isopods are the main dietary elements. On the Ogasawara Islands, it appears to be displaced by the invasive *Anolis carolinensis*, which contests it for prey.

Cryptoblepharus novohebridicus　　　　　　**Vanuatu Snake-eyed Skink**
Plate 16

APPEARANCE: Small, slightly elongate-bodied lizard with short, well-developed limbs and long, subcylindrical tapering tail. Its head is ovate in dorsal outline and barely distinct from neck. Eyes lack moveable eyelids, and as with snakes, each eye is protected by clear scale. Dorsally, broad, medium- to dark-brown stripe extends from snout to base of tail; this dorsal stripe is bordered on each side by narrow, black-edged white dorsolateral stripe from eye to tail; broad, dark-brown to black lateral stripe from loreal area onto tail; and below white stripe from upper lip to hindlimb. Tail bears series of transverse black marks on blue background. Venter is bluish white, usually unmarked. Neck, trunk, and tail have large smooth, imbricate scales above and below; middorsal pairs are enlarged, about one and a half times larger than lateral and ventral scales. Scales on head are smooth, enlarged abutting plates; supranasal and interparietal scales are absent; there are typically seven supralabials, fifth one is largest and beneath eye. First two pairs of middorsal body scales behind parietals are greatly enlarged as nuchal plates. Middorsal scales are usually in 49 to 58 rows from nape to base of tail, 21 to 26 scale rows encircle body in middle of trunk, and fourth toe (hindfoot) bears 18 to 23 lamellae on its underside.

SIZE: Adult females appear to average slightly larger than males. Adult females range from 32 to 37 mm SVL (mean, 35 mm); males from 29 to 36 mm (33 mm). Hatchlings are about 20 mm SVL. Tail length is 125% to 140% of SVL.

OCCURRENCE: This skink is endemic to Vanuatu, although its presence has not been confirmed for about half of the larger islands.

HABITAT CHOICE: *C. novohebridicus* is mainly a shoreline inhabitant and most common in rocky areas.

REPRODUCTION: The reproductive biology of this species has not been studied, although gravid females carry one to two eggs.

MISCELLANEA: Daily activity and prey is probably the same as the Oceania Snake-eyed Skink.

Cryptoblepharus poecilopleurus   **Oceania Snake-eyed Skink**
Plate 16

APPEARANCE: Small to midsize, slightly elongate-bodied lizard with short, well-developed limbs and long, subcylindrical tapering tail. Its head is ovate in dorsal outline and barely distinct from neck. Eyes lack moveable eyelids, and as with snakes, each eye is protected by clear scale. Dorsal ground color is variable, although consistent within local populations, from golden tan to bright coppery brown. Ground color forms broad dorsal stripe from snout onto tail; stripe is commonly densely marked with dark-brown or black flecks and bordered by cream to tan dorsolateral stripe from eye onto tail; this stripe is roughly edged by black both above and below; lower edge spreads ventrally on trunk and bears numerous small cream spots. Tail is typically lighter shade of ground color. Venter ranges from bluish white to creamy tan, usually unmarked. Neck, trunk, and tail have large smooth, imbricate scales above and below; middorsal pairs are enlarged, about one and a half times larger than lateral and ventral scales. Scales on head are smooth, enlarged, abutting plates; supranasal and interparietal scales are absent; there are typically seven supralabials, fifth one is largest and beneath eye. First two pairs of middorsal body scales behind parietals are greatly enlarged as nuchal plates. Middorsal scales are usually in 49 to 60 rows from nape to base of tail, 25 to 30 scale rows encircle body in middle of trunk, and fourth toe (hindfoot) bears 14 to 24 lamellae on its underside.

SIZE: Adult females and males are equal in size. Adult females range from 37 to 51 mm SVL; males from 37 to 48 mm SVL. There appear to be slight differences in body size among the widespread populations. Tail length is 135% to 150% of SVL. Hatchling size ranges from 21 to 23 mm SVL.

OCCURRENCE: This skink is widespread, occurring from the Mariana Islands and Hawaiian Islands in the north to Rapa Nui (Easter Island) in the south, and from western Micronesia and Tonga eastward to the South American mainland (Ecuador and Peru), although it does not appear in the Galápagos Islands or other eastern Pacific islands.

HABITAT CHOICE: *C. poecilopleurus* is largely a beach inhabitant. It occupies the typical beach habitats of mangroves, rock shorelines, and beachside trees and brush tangles. Like the other Pacific *Cryptoblepharus*, it appears to require structured sites above the high-tide line that possess crevices and similar, solid-walled resting and likely nesting sites.

REPRODUCTION: *C. poecilopleurus* females lay one or two eggs at a time; in the American Samoa population, most females bore two eggs. The eggs are seldom found, hence the assumption of deposition sites deep in crevices in rocks and other solid structures, including the dense root masses of palms. A communal nest of over 70 hatched and unhatched eggs was found in moist soil adjacent to a cattle-guard in Maui; several of the eggs hatched when found, confirming their specific identity. In the Marianas, a reproductive study showed reproductive activity year-round and the likelihood that females lay multiple times each year. This aseasonal reproduction is likely for other Pacific populations.

MISCELLANEA: Snake-eyed Skinks become active as soon as sunlight hits and begins to warm their rocky abodes. They emerge from their resting cracks and crevices, bask briefly (presumably elevating their body temperatures), and begin to forage. Small invertebrates, small isopods, and amphipods are probably their major prey. They usually are active throughout the day, but data are not available to identify whether an individual has this prolonged activity or individuals intermittently and alternately forage and rest.

Cryptoblepharus rutilus   **Palau Snake-eyed Skink**

Plate 16

APPEARANCE: Small, slightly elongate-bodied lizard with short, well-developed limbs and long, subcylindrical tapering tail. Its head is ovate in dorsal outline and barely distinct from neck. Eyes lack moveable eyelids, and as with snakes, each eye is protected by clear scale. A broad stripe of ground color extends from snout onto tail, generally silvery tan; stripe is usually strongly marked with small, dark-gray blotches on neck and trunk, smudgelike on head; dorsal stripe is bordered on each side by narrow white to cream dorsolateral stripe from tip of snout above eye onto tail, bordered above and below by broad, ragged black edges; broad dark-brown to black lateral stripe from head onto base of tail, and below narrow white to cream from upper lip beneath eye to hindlimb. Tail is distinctly lighter than trunk dorsally. Venter is shiny bluish ivory from chin to vent, yellowish tan on underside of tail. Neck, trunk, and tail have large smooth, imbricate scales above and below; middorsal pairs are enlarged, about one and a half times larger than lateral and ventral scales. Scales on head are smooth, enlarged, abutting plates; supranasal and interparietal scales are absent; there are typically seven supralabials, fifth one is largest and beneath eye. First two to three pairs of middorsal body scales behind parietals are greatly enlarged as nuchal plates. Middorsal scales are usually in 45 to 51 rows from nape to base of tail, 20 to 24 scale rows encircle body in middle of trunk, and fourth toe (hindfoot) bears 21 to 24 lamellae on its underside.

SIZE: Adult females and males are equal in size. Adult females and range from 34 to 38 mm SVL (mean, 36 mm); males from 33 to 36 mm (35 mm). Hatchlings are about 20 mm SVL.

OCCURRENCE: This skink is endemic to the Palau Islands and occurs on most of the larger islands.

HABITAT CHOICE: *C. rutilus* is both a beach and an inland inhabitant. It occupies the typical beach habitats of mangroves, rock shorelines, beach-side trees, and brush tangles; however, it also regularly lives inland distant from beaches in forests.

REPRODUCTION: Reproductive data are not available. Presumably, most females bear a single egg.

Emoia adspersa   **Striped Small-scaled Skink**

Plate 23, Figure 14

APPEARANCE: Midsize, robust lizard with long limbs and moderately long tail. Its head is ovate in dorsal outline and distinct from neck. Dorsal ground color is uniform shiny coppery brown to orangish brown, occasionally dark brown; numerous small, dark-brown blotches are scattered over dorsum and sides from neck onto tail; occasionally dark lateral stripe from loreal area to forelimb and few light spots can be present on side of trunk. Head color matches body and immaculate dorsally; lips are lighter brown. Venter from chin onto trunk is unicolor white or cream to light tan. Neck, trunk, and tail have small, smooth, imbricate scales; scales are nearly equal in size around body. Scales on head are smooth, enlarged abutting plates; supranasal scales are present, interparietal scale is moderate sized with parietal eye; anterior loreal is higher than long, and there are usually eight supralabials with sixth largest and beneath eye. First pair of middorsal body scales behind parietals is enlarged as nuchal plates. Middorsal scales are in 93 to 114 rows from nape to base of tail, 50 to 60 scale rows encircle body in middle of trunk, and fourth toe (hindfoot) bears 23 to 31 broad, smooth lamellae on its underside.

SIZE: Adult females and males are equal in size. Adult females range from 63 to 81 mm SVL; males from 64 to 85 mm SVL. Tail length is 115% to 140% of SVL. Hatchling size has not been reported.

OCCURRENCE: *E. adspersa* is a central Pacific resident, living on Funafuti (Tuvalu), Nukunonu (Tokelau), Futuna, throughout the Samoan islands, and Niuafo'ou (Tonga).

HABITAT CHOICE: This skink is a terrestrial to semi-arboreal resident of heavily vegetated shorelines into the open coastal woodlands.

REPRODUCTION: Females typically produce clutches of two eggs, occasionally a single egg. Eggs have not been located in the wild, but they likely are deposited beneath forest-floor litter, rocks, and logs.

Emoia aneityumensis   **Anatom Treeskink**

Plate 19

APPEARANCE: Midsize, medium-bodied lizard with moderately long limbs and long, tapering tail. Its head is ovate in dorsal outline and distinct from neck. Dorsal ground color is light brown with faint coppery overtones; dorsum is nearly uniform and lighter brown on head; small, dark-brown blotches form a dorsolateral series; numerous pale spots cover trunk between limbs and onto tail; limbs dorsally are golden tan with pale spots and few small, black marks. Venter is creamy white. Neck, trunk, and tail have moderate-sized, smooth, imbricate scales; scales are nearly equal in size around body. Scales on head are smooth, enlarged abutting plates; supranasal scales are present, and interparietal scale is modest in size with parietal eye; anterior loreal is roughly square, and there are usually seven supralabials with either fifth or sixth largest and beneath eye. First pair of middorsal body scales behind parietals is enlarged as nuchal plates. Middorsal scales are in 74 to 80 rows from nape to base of tail, 39 to 41 scale rows encircle body in middle of trunk, and fourth toe (hindfoot) bears 57 to 71 broad, smooth lamellae on its underside.

SIZE: Adult females might be smaller than males; however, sample sizes are inadequate to test the differences. Adult females range from 71 to 88 mm SVL; males from 90 to 96 mm SVL. Tail length is 140% to 170% of SVL.

OCCURRENCE: *E. aneityumensis* lives only on Anatom Island in southern Vanuatu.

HABITAT CHOICE: This largely arboreal species lives in open-canopied forests.

REPRODUCTION: Females bear clutches of four to five eggs. Additional reproductive data are absent, although females probably descend to the ground and lay their eggs in the forest-floor litter adjacent to the base of a tree.

Emoia arnoensis   **Micronesia Black Skink**
Plate 22

APPEARANCE: Midsize, robust lizard with moderately long limbs and long tapering tail. Its head is ovate in dorsal outline and distinct from neck. Dorsal ground color is uniform shiny black. Head matches body coloration. Venter from chin onto trunk is uniform dark brown, slightly lighter than dorsal color. Neck, trunk, and tail have moderate-sized, smooth, imbricate scales; scales are nearly equal in size around body. Scales on head are smooth, enlarged abutting plates; supranasal scales are present; interparietal scale is modest in size with parietal eye; anterior loreal is slightly higher than long, and there are usually eight supralabials with sixth largest and beneath eye. First pair of middorsal body scales behind parietals is enlarged as nuchal plates. Middorsal scales are in 66 to 75 rows from nape to base of tail, 36 to 40 scale rows encircle body in middle of trunk, and fourth toe (hindfoot) bears 35 to 42 broad, smooth lamellae on its underside.

SIZE: Males average slightly larger than females. Adult females range from 74 to 83 mm SVL; males from 73 to 86 mm SVL. Tail length is about 140% to 170% of SVL.

HABITAT CHOICE: This skink is strongly terrestrial, typically living in open habitats that are often thickly weeded or vine-covered from forest edge through fields and gardens to yards.

OCCURRENCE: *E. arnoensis* is an eastern Micronesian species, residing in Kosrae, Marshall Islands, and Nauru.

REPRODUCTION: Gravid females bear two eggs, one in each oviduct. Other reproductive data are not available.

MISCELLANEA: The Nauru population has been described as a distinct subspecies from the Micronesian-Marshall Islands populations. Nauru skinks are similar in most traits with the latter population, with the exception of possessing more middorsal scales (82 to 87). A record of *E. arnonensis* from Chuuk appears to result from an incorrectly labeled specimen.

E. *arnoensis* is a member of the *Emoia atrocostata* group, a moderately diverse species assemblage of coastal and insular species distributed from eastern Malaysia and Christmas Island eastward across the Lesser Sunda Islands into the central Pacific.

Emoia atrocostata   **Seaside Skink**

Plate 23

APPEARANCE: Midsize, medium-bodied lizard with modest length limbs and long, tapering tail. Its head is ovate in dorsal outline and slightly distinct from neck. Dorsally, body is two-toned, often with shades of gray to brown on the back and dark gray or brown to nearly black on the sides of trunk. These background colors are commonly overlain with variable pattern of darker and/or light marks. Head matches body coloration, with lips lighter shade. Venter from chin onto trunk is uniform light gray to creamy ivory. Neck, trunk, and tail have moderate-sized, smooth, imbricate scales; scales are nearly equal in size around body. Scales on head are smooth, enlarged, abutting plates; supranasal scales are present, and interparietal scale is small, elongate pentagonal shape with parietal eye; anterior loreal is higher than long, and there are usually eight supralabials with sixth largest and beneath eye. First pair of middorsal body scales behind parietals is enlarged as nuchal plates. Middorsal scales are in 61 to 76 rows from nape to base of tail, 32 to 43 scale rows encircle body in middle of trunk, and fourth toe (hindfoot) bears 30 to 42 broad, smooth lamellae on its underside.

SIZE: Males average slightly larger than females in some populations. Adult females range from 57 to 92 mm SVL; males from 60 to 98 mm SVL. Tail length is 140% to 180% of SVL. Hatchlings are 33 to 39 mm SVL for Philippine population, and 20 to 25 mm SVL for Guam population.

OCCURRENCE: Within the tropical Pacific, *E. atrocostata* inhabits the western band of islands from the Santa Cruz Islands and Vanuatu northward through Micronesia into the northern Marianas. Its eastward limits are in Pohnpei. It also occurs coastally from Malaysia through Greater and Lesser Sunda Islands to New Guinea and Cape York, Australia, and eastward into the Philippines, Taiwan, and the Solomon Islands.

HABITAT CHOICE: This skink is terrestrial and seldom found far from rocky seashores. It actively forages among the rocks, readily leaping from one to another and even diving into the water to hide from disturbances. It also regularly lives in mangrove swamps.

REPRODUCTION: Gravid females typically bear two eggs, one in each oviduct. They deposit their eggs in tree holes and the cracks and crevices of their rocky residences. This species is suspected of reproducing year-round in the Philippines, but other populations such as the one in Lanyu (Taiwan) have a seasonal reproductive cycle, with the females reproductively active March through August. Lanyu female clutches average two eggs, ranging from one to three.

MISCELLANEA: Three subspecies are recognized: one in the tropical Pacific and most of the eastern Sundan and New Guinea area; a second on Cape York; and a third in the Solomon Islands and Vanuatu. Owing to geographical variable, color patterns, and differences in average adult sizes, it is highly likely that today's *E. atrocostata* will become multiple species after detailed analysis of morphological and molecular variations.

   E. atrocostata is the most widely occurring member of the *Emoia atrocostata* group. The range of *E. atrocostata* encompasses that of all other group members except for the outlying *Emoia laobaoense* that live on an isolated mountain range in Vietnam.

Emoia boettgeri   **Micronesia Spotted Skink**

Plate 22

APPEARANCE: Midsize, medium-bodied skink with moderately long limbs and long, tapering tail. Its head is ovate in dorsal outline and distinct from neck. Dorsally, ground color is olive to brown, either uniformly colored or more commonly spotted with either numerous indistinct small, dark spots or larger, light spots. Venter from chin onto trunk is light gray, usually lighter on chin. Neck, trunk, and tail have moderate-sized, smooth, imbricate scales; scales are nearly equal in size around body. Scales on head are smooth, enlarged, abutting plates; supranasal scales are present, and interparietal scale is moderately large with parietal eye; anterior loreal is higher than long or height and length are nearly equal, and there are usually eight supralabials with sixth largest and beneath eye. First pair of middorsal body scales behind parietals is enlarged as nuchal plates. Middorsal scales are in 60 to 72 rows from nape to base of tail, 36 to 42 scale rows encircle body in middle of trunk, and fourth toe (hindfoot) bears 42 to 52 broad, smooth lamellae on its underside.

SIZE: Males and females are equal in size. Adult females range from 62 to 77 mm SVL; males from 60 to 77 mm SVL. Tail length is 180% to 220% of SVL.

OCCURRENCE: *E. boettgeri* is a central and eastern Micronesian species, residing in Chuuk eastward to Arno Atoll, Marshall Islands.

HABITAT CHOICE: This skink is strongly terrestrial, and typically resides in open habitats from the forest edge through fields and gardens to yards.

REPRODUCTION: Females probably bear two eggs. Reproductive data are not available.

Emoia caeruleocauda   **Pacific Blue-tailed Skink**

Plate 17, Figure 15

These skinks are the true blue-tailed lizards of the Pacific. Other *Emoia* have bluish tails, but none can outclass the eye-catching cerulean blue, tail-wagging juveniles of *E. caerueleocauda*. The constant undulation of foraging juveniles gives credence to a brightly colored tail serving as a focus for a predator's attack.

APPEARANCE: Midsize, medium-bodied lizard with moderately long limbs and long, tapering tail. Its head is ovate in dorsal outline and distinct from neck. Dorsal ground color is glossy dark brown to black and bears three to five white, body-length stripes. Middorsal stripe extends from tip of snout to base of tail; pair of dorsolateral stripes are present from above eye onto tail, and a lateral stripe on each side from posterior edge of eye to hindlimb. Tail is blue above and below, brightest blue in juveniles. Occasionally, individuals lack stripes and are uniformly dark above. Venter is unicolor white to ivory. Neck, trunk, and tail have moderate-sized, smooth, imbricate scales; scales are nearly equal in size around body. Scales on head are smooth, enlarged, abutting plates; supranasal scales are present, and interparietal scale and

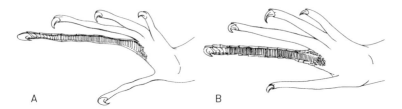

FIGURE 15. Contrasting digital lamellar morphology in blue-tailed *Emoia*.
E. caeruleocauda (A) has broad and fewer lamellae than *E. impar* (B) and
E. cyanura, which have narrow blade-like lamellae.

interparietal eye are absent; anterior loreal is higher than long, and there are usually six to seven supralabials with fifth largest and beneath eye. First pair of middorsal body scales behind parietals is enlarged as nuchal plates. Middorsal scales are in 50 to 64 rows from nape to base of tail, 27 to 36 scale rows encircle body in middle of trunk, and fourth toe (hindfoot) bears 33 to 54 broad, smooth lamellae on its underside. **SIZE:** Males average slightly larger than females. Adult females range from 41 to 55 mm SVL; males from 40 to 66 mm SVL. Tail length is ~150 % of SVL. Hatchlings range from about 22 to 24 mm SVL. **OCCURRENCE:** As presently defined, *E. caeruleocauda* is the most broadly distributed Pacific skink. Its range extends from the Mollucas and Celebes through New Guinea and its associated islands westward to Fiji and northward through Micronesia to the Northern Mariana Islands, and includes Palau and some Philippine islands. **HABITAT CHOICE:** *E. caeruleocauda* is predominantly a forest-edge inhabitant foraging low on the branches of shrubs, understory trees, and bramble tangles. **REPRODUCTION:** Like many widespread Pacific skinks, the reproductive biology of *E. caeruleocauda* is poorly studied. Females usually produce two eggs, infrequently a single egg, that are laid within and beneath the forest-floor litter; communal nest have been found. In Vanuatu, gravid females occur year-round with the highest frequency between September and February. Incubation duration is about 26 days. **MISCELLANEA:** Walter Brown, in his *Emoia* monograph entitled "*E. caeruleocauda*: A Superspecies," indicates that the present taxon is a group of cryptic or taxonomically undifferentiated species. The broad distribution of *E. caeruleocauda* supports this concept. No researcher has identified morphological features that can serve to segregate multiple species; however, it is likely the western (New Guinea area, possibly eastward into Vanuatu) populations are distinct from the Mariana ones, and perhaps populations in other areas have speciated as well.

Emoia campbelli   Vitilevu Mountain Treeskink
Plate 21

APPEARANCE: Midsize, medium-bodied lizard with moderately long limbs and long, tapering tail. Its head is ovate in dorsal outline and distinct from neck. Dorsal ground color is medium brown with coppery overtone, densely overlain with medium-sized creamy yellow spots on back and upper half of sides and small, dark brown blotches throughout; head shares dorsal ground color and scattering of smaller, dark blotches. Venter from chin onto trunk is immaculate yellow to greenish yellow with reddish tint on sacrum and base of tail. Neck, trunk, and tail have moderate-sized, smooth, imbricate scales; scales are nearly equal in size around body. Scales on head are smooth, enlarged, abutting plates; supranasal scales are present, interparietal scale is small; anterior loreal is longer than high, and there are usually eight supralabials with sixth largest and beneath eye. First pair of middorsal body scales behind parietals is enlarged as nuchal plates. Middorsal scales are in 56 to 64 rows from nape to base of tail, 30 to 36 scale rows encircle body in middle of trunk, and fourth toe (hindfoot) bears 48 to 54 broad, smooth lamellae on its underside.

SIZE: Females and males are equal in size. Adult females range from 68 to 98 mm SVL; males from 70 to 98 mm SVL. Tail length is ~150% of SVL. Hatchling size is unknown.

OCCURRENCE: *E. campbelli* lives in the mountains of central Viti Levu on the Rairaimatuka Plateau.

HABITAT CHOICE: This skink is a forest inhabitant, usually seen on trees and occasionally on smaller understory trees.

REPRODUCTION: Females lay two eggs, and they regularly, if not exclusively, deposit their eggs in the chambers of the epiphytic ant plant.

Emoia concolor   Fiji Slender Treeskink
Plate 21

APPEARANCE: Midsize, slender to medium-bodied lizard with moderately long limbs and long, tapering tail. Its head is ovate in dorsal outline and distinct from neck. Dorsal ground color is variable, ranging from uniform light green or green anterior and coppery tan posterior to uniform coppery brown, occasionally with widely scattered small, dark spots and less commonly with white spots. Head color matches body color and usually is unicolor dorsally; lips and loreal are lighter. Venter from chin onto trunk is unicolor yellowish white to lime green. Neck, trunk, and tail have moderate-sized, smooth, imbricate scales; scales are nearly equal in size around body. Scales on head are smooth, enlarged, abutting plates; supranasal scales are present, and interparietal scale is small, elongate pentagonal, with parietal eye; anterior loreal is longer than high, and there are usually eight supralabials with sixth largest and beneath eye. First pair of middorsal body scales behind parietals is enlarged as nuchal plates. Middorsal scales are in 54 to 62 rows from nape to base of tail, 27 to 33 scale rows encircle body in middle of trunk, and fourth toe (hindfoot) bears 30 to 48 broad, smooth lamellae on its underside.

SIZE: Males average larger than females. Adult females range from 58 to 80 mm SVL; males from 62 to 95 mm SVL. Tail length is 150% to 200% of total length. Hatchlings range from 26 to 28 mm SVL.

OCCURRENCE: *E. concolor* is a Fijian endemic and occurs throughout the Fijian islands, from the smallest sand island to the large basaltic ones. Some of the color variation is regional, which probably reflects the presence of cryptic species among the Fijian populations of *E. concolor*.

HABITAT CHOICE: This treeskink is strongly arboreal and uncommonly found on the ground. It lives in the shrubbery and trees of urban landscapes to the forest edge and broken-canopy forests.

REPRODUCTION: Females bear clutches of one to four eggs. They descend to the ground for egg-laying, regularly depositing their eggs shallowly in the soil beneath leaf litter at the base of trees.

Emoia cyanogaster   **Green-bellied Vineskink**

Plate 21

APPEARANCE: Midsize, elongate, medium-bodied lizard with moderately short limbs and very long, tapering tail. Its head is elongate ovate in dorsal outline and distinct from neck. Dorsal ground color is medium brown, strongly coppery on head and anterior neck and losing shiny appearance on trunk and tail, immaculate on head to midtrunk with scattering of small light and dark spots thereafter; broad, dark lateral stripes extend from loreal area to midtrunk, providing demarcation between brown back and light-green venter. Neck, trunk, and tail have moderate-sized, smooth, imbricate scales; middorsal and adjacent scales are moderately large with gradual size decrease to uniformly sized ventral scales. Scales on head are smooth, enlarged abutting plates; supra-nasal scales are present, interparietal scale is moderate sized, diamond shaped with parietal eye; anterior loreal is longer than high, and there are usually eight supralabials with sixth largest and beneath eye. First pair of middorsal body scales behind parietals is enlarged as nuchal plates. Middorsal scales are in 54 to 62 rows from nape to base of tail, 22 to 28 scale rows encircle body in middle of trunk, and fourth toe (hindfoot) bears 70 to 90 very narrow, smooth lamellae on its underside.

SIZE: Females and males are equal in size. Adult females range from 68 to 90 mm SVL; males from 62 to 92 mm SVL. Tail length is nearly 200% of SVL. Hatchling size is unknown.

OCCURRENCE: *E. cyanogaster* occurs from the islands of the Bismarck Archipelago eastward through the Solomon Islands to the northern island group of Vanuatu.

HABITAT CHOICE: This skink is strongly arboreal, living in vine and brush tangles at forest edges and abandoned gardens. It is an active forager, readily moving across thin branches and vines in search of insect and spider prey.

REPRODUCTION: Females bear two eggs and deposit them in loose soil beneath leaf and other plant debris. Other aspects of its reproductive biology are not documented.

Emoia cyanura　　　　　　　　　　**White-bellied Copper-striped Skink**

Plate 17, Figures 15, 16

APPEARANCE: Midsize, medium-bodied lizard with moderately long limbs and long, tapering tail. Its head is ovate in dorsal outline and distinct from neck. Dorsal ground color is black to dark copper, overlain by three bright, copper-colored stripes; middorsal stripe extends from tip of snout onto base of tail; pair of dorsolateral stripes from above eyes to sacrum, although commonly fading at midtrunk in adults. Tail is bluish in some small juveniles, typically greenish brown in juveniles and adults, often brown in adults. Venter from chin onto trunk is uniform shiny white or white with coppery overtone. Neck, trunk, and tail have moderate-sized, smooth, imbricate scales; scales are nearly equal in size around body, and middorsal paired row of scales never fused into larger single scales. Scales on head are smooth, enlarged, abutting plates; supranasal scales are present, and interparietal scale is absent, distinct parietal eye in posteromedial tip of frontoparietal scale; anterior loreal is longer than high or nearly rectangular, and there are usually seven supralabials with fifth largest and beneath eye. First pair of middorsal body scales behind parietals is enlarged as nuchal plates. Middorsal scales are in 53 to 63 rows from nape to base of the tail, 28 to 34 scale rows encircle body in the middle of trunk, and fourth toe (hindfoot) bears 36 to 51 (rarely less than 40) narrow, smooth lamellae on its underside.

SIZE: Males average slightly larger than females. Adult females range from 39 to 53 mm SVL (mean, 46 mm); males from 39 to 56 mm SVL (48 mm). Tail length is 150% to 175% of SVL. Hatchlings range from 20 to 23 mm SVL.

OCCURRENCE: *E. cyanura* occurs widely, but not everwhere, in Oceania, from the Solomon Islands westward into French Polynesia and northward to eastern Micronesia (Guam and Wake). The Hawaiian population was a recent introduction to the grounds of a Kaua'i hotel and now is extirpated.

HABITAT CHOICE: The White-bellied Copper-striped Skink is predominantly a terrestrial species, although occasionally it forages low on the base and in lower branches of trees or axils of palms and pandanus. It is an inhabitant of open-sky situations from the lawns and shrubbery of urban landscapes to grassy beachsides; gardens; and open-canopied, secondary-growth forest.

REPRODUCTION: Females lay two egg clutches, rarely a single egg, usually in a shallow nest beneath plant debris, rocks, and logs. When eggs are found, the site often harbors the egg clutches of more than one female is an example of communal nesting. The Tahiti population lays eggs year-round, with peak reproductive activity between October and March; females were gravid year-round also in American Samoa. This aseasonal reproductive activity is likely for most populations. Incubation probably averages between 40 and 50 days.

Emoia erronan   **Erronan Treeskink**

Plate 19

APPEARANCE: Midsize, robust lizard with moderately long limbs and long, tapering tail. Its head is ovate in dorsal outline and distinct from neck. Head and body ground colors are in striking contrast and two colorations exist: bright, reddish-copper head and neck with medium to dark shiny brown body, limbs, and tail; or head and neck coppery white with black body, limbs, and tail. Venter in both color morphs is light to dusky. Neck, trunk, and tail have moderate-sized, smooth, imbricate scales; scales are nearly equal-sized around body. Scales on head are smooth, enlarged

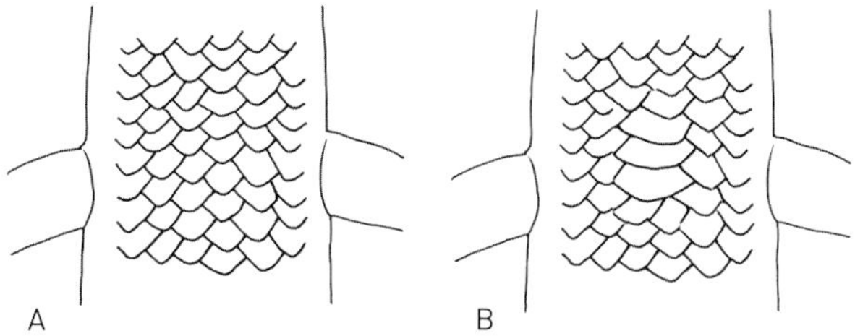

FIGURE 16. Dorsal scalation morphology of striped *Emoia*. *E. cyanura* (A) has a continuous row of paired middorsal scales from nape on to tail; this condition is typical for most *Emoia* skinks. *E. impar* (B) has some or most of the middorsal scale pairs fused into single median scales.

abutting plates; supranasal scales are present, and interparietal scale is small, slender, with parietal eye; anterior loreal is higher than long, and there are usually eight supralabials with sixth largest and beneath eye. First pair of middorsal body scales behind parietals is enlarged as nuchal plates. Middorsal scales are in 75 to 84 rows from nape to base of tail, 36 to 41 scale rows encircle body in middle of trunk, and fourth toe (hindfoot) bears 37 to 51 smooth lamellae on its underside.

SIZE: Males may average slightly larger than females. Adult female size is presently unknown; adult males range from 69 to 101 mm SVL (mean, 85 mm). Tail length is 135% to 170% of SVL. Hatchling size is unknown.

OCCURRENCE: *E. erronan* lives only on Futuna Island, Vanuatu.

HABITAT CHOICE: Habitat preference has not been reported. Presumably, it is a predominantly arboreal lizard foraging in the trees and uncommonly on the ground.

REPRODUCTION: No reproductive data are available; presumably females lay mainly two-egg clutches.

Emoia impar   **Dark-bellied Copper-striped Skink**

Plate 17, Figure 16

APPEARANCE: Midsize, medium-bodied lizard with moderately long limbs and long, tapering tail. Its head is ovate in dorsal outline and distinct from neck. Dorsal ground color is black to dark copper, overlain by three bright, copper-colored stripes, usually distinct to anterior trunk and often fading thereafter; middorsal stripe extends from tip of snout to tail; pair of dorsolateral stripes from above eyes to sacrum; in some, presumably older individuals, stripes are largely lost and skink is entirely dark copper. Tail is bright blue in juveniles, fading in adults although remaining distinctly blue. Venter is uniformly dusky from chin onto tail. Neck, trunk, and tail have moderate-sized, smooth, imbricate scales; scales are nearly equal in size around body, and middorsal paired row of scales almost always has some pairs fused into larger, single scales. Scales on head are smooth, enlarged abutting plates; supranasal scales are present, and interparietal scale is absent, parietal eye is absent; anterior loreal is higher than long, and there are usually seven supralabials with fifth largest and beneath eye. First pair of middorsal body scales behind parietals is enlarged as nuchal plates. Middorsal scales are in 52 to 61 rows from nape to base of tail, 27 to 33 scale rows encircle body in middle of trunk, and fourth toe (hindfoot) bears 59 to 80 very narrow, smooth lamellae on its underside.

SIZE: Females and males are similar in size. Adult females range from 40 to 47 mm SVL (mean, 45 mm); males from 40 to 47 mm SVL (45 mm). Tail length is 125% to 175% of SVL. Hatchlings range from 22 to 23 mm SVL.

OCCURRENCE: This blue-tailed skink has a broad distribution throughout the Pacific, largely overlapping the range of *E. cyanura*. The range similarly extends eastward from the Bismarck Archipelago to French Polynesia, but it is less northerly in eastern Micronesia. It was a resident of most of the Hawaiian Islands, but recent surveys suggest that it is now extinct there, although a recent sight record of a blue-tailed lizard on an islet off the Moloka'i coast suggests that it may persist. Its extinction is probably more the result of the presence of an invasive ant rather than competitive displacement by *Lamprolepis delicata*. Whether this skink arrived in the Hawaiian Islands with the Polynesians or before them remains uncertain.

HABITAT CHOICE: *E. impar* is largely a shade-loving species and occurs mainly in strong canopied forests or at their edges. Like *E. cyanura*, it is predominantly a terrestrial species, although it forages occasionally low on the base of trees and in the lower branches of trees or axils of palms and pandanus.

REPRODUCTION: Females lay clutches with typically two eggs, rarely one, usually in shallow nests beneath plant debris and logs. The Tahiti population lays year-round, with peak reproductive activity between October and March. In one instance, an old coconut shell was found on the ground with over 400 *E. impar* eggs inside, 136 of which contained embryos.

Emoia jakati   **Papua Five-striped Skink**

Plate 18

APPEARANCE: Midsize, medium-bodied lizard with moderately long limbs and long, tapering tail. Its head is ovate in dorsal outline and distinct from neck. Overall dorsal coloration is four light stripes on dark background. Middorsally, medium-brown ground color extends from snout onto tail; dorsolaterally on each side, white stripe extends from anterior neck to posterior trunk, occasionally to base of tail; another white stripe extends from upper lip beneath eye to hindlimb; these two pairs of white stripes are embedded in dark-brown lateral ground color. Venter is unicolor light gray from chin to white tail. Neck, trunk, and tail have moderate-sized, smooth, imbricate scales; scales are nearly equal in size around body. Scales on head are smooth, enlarged, abutting plates; supranasal scales are present, and interparietal and parietal eyes are absent; anterior loreal is higher than long, and there are usually eight supralabials with sixth largest and beneath eye. First pair of middorsal body scales behind parietals is enlarged as nuchal plates. Middorsal scales are in 60 to 72 rows from nape to base of tail, 29 to 39 scale rows encircle body in middle of trunk, and fourth toe (hindfoot) bears 30 to 42 broad, smooth lamellae on its underside.

SIZE: Females and males are equal in size. Adult females range from 43 to 53 mm SVL; males from 37 to 51 mm SVL. Tail length is 150% to 170% of SVL. Hatchling size is unknown.

OCCURRENCE: *E. jakati* is a New Guinean skink, occurring from Ceram through northern New Guinea and the Bismarck Archipelago into the Solomon Islands. In Oceania, it has a spotty distribution in Palau, the Federated States of Micronesia (Yap, Chuuk, Pohnpei, and Kosrae) and a few atolls of the Marshall Islands. When and how *E. jakati* arrived in these Oceania localities is unknown, although most of its locality precedes World War II interisland transport of war materials. The Oceania localities seem to match late-19th-century German trading routes.

HABITAT CHOICE: In Oceania, *E. jakati* lives mainly in highly disturbed areas within villages and town to gardens and degraded secondary forest areas.

REPRODUCTION: Reproductive data are limited. Females bear clutches of two eggs.

MISCELLANEA: *E. jakati* is considered to be a complex of multiple species, all morphologically similar, although it seems likely that the various Oceania populations derived from a single New Guinean population.

Emoia lawesii   **Olive Small-scaled Skink**

Plate 23

APPEARANCE: Midsize, robust lizard with moderately short limbs and medium-length, tapering tail. Its head is elongate ovate in dorsal outline and distinct from neck. Dorsal ground color ranges from tan to dark brown; head is typically immaculate with marking appearing at midneck and extending onto tail. Lighter backgrounds show some dark-brown marks or mottling and often some yellowish-white spots; as background darkens, the dark mottling becomes heavier; limb color matches body or slightly lighter. Venter from chin onto tail is immaculate creamy white to beige. Neck, trunk, and tail have small, smooth, imbricate scales; scales are nearly equal in size around body. Scales on head are smooth, enlarged, abutting plates; supranasal scales are present, and interparietal scale is modest, elongate pentagonal shape with parietal eye; anterior loreal is higher than long, and there are usually eight supralabials with sixth largest and beneath eye. First pair of middorsal body scales behind parietals is enlarged as nuchal plates. Middorsal scales are in 99 to 112 rows from nape to base of tail, 48 to 62 scale rows encircle body in middle of trunk, and fourth toe (hindfoot) bears 27 to 34 broad, smooth lamellae on its underside.

SIZE: Females and males are equal in size. Adult females range from 78 to 105 mm SVL; males from 77 to 106 mm SVL. Tail length is approximately 100% to 120% of SVL. Hatchlings range from 32 to 33 mm SVL.

OCCURRENCE: *E. lawesii* has a small distribution among the south-central Pacific islands, present on a few islands (Auʻnuu, Olosega, Taʻu) in Samoa and Niue. A voucher specimen in the British natural history museum is purported to be from Tongatapu. Recent surveys on that Tongan island and others have not found this skink. Because it is a distant outlier from other *E. lawesii* population, this record is either an error or a stowaway.

HABITAT CHOICE: This small-scaled skink is predominantly a terrestrial forager of forest-beach edges and open-canopied forests.

REPRODUCTION: Females typically produce clutches of two eggs, occasionally a single egg. Eggs have not been located in the wild, but they are probably deposited beneath forest-floor litter, beneath rocks and logs, and in deep crevices and holes in coral limestone. Incubation is long, about 70 to 78 days. Year-round reproduction is possible, although not confirmed.

Emoia mokolahi   **Tonga Robust Treeskink**
Plate 20

APPEARANCE: Midsize, robust lizard with moderately long limbs and long, tapering tail. Its head is ovate in dorsal outline and distinct from neck. Dorsal ground color is light to medium brown, coppery on head, less shiny on neck and trunk; series of short, longitudinal dark-edged streaks of cream and infrequent small, dark-brown spots from midneck to base of tail; laterally three modest-sized dark-brown spots, one on neck, another above shoulder, and final one on anterior trunk. Head matches body coloration. Venter is cream colored with orange highlights, chin to anterior chest immaculate, dark speckling begins behind axilla and continues onto base of tail. Undersides of limbs lighter colored than venter, and underside of fore- and hindfeet are bright yellow. Neck, trunk, and tail have moderate-sized, smooth, imbricate scales; scales are nearly equal in size around body. Scales on head are smooth, enlarged, abutting plates; supranasal scales are present, and interparietal scale is modest-sized and diamond-shaped, with parietal eye; anterior loreal is higher than long, and there are usually eight supralabials with sixth largest and beneath eye. First pair of middorsal body scales behind parietals is enlarged as nuchal plates. Middorsal scales are in 58 to 72 rows from nape to base of tail, 30 to 36 scale rows encircle body in middle of trunk, and fourth toe (hindfoot) bears 42 to 51 broad, smooth lamellae on its underside.

SIZE: Females and males are equal in size. Adult females range from 86 to 105 mm SVL (mean, 96 mm); males from 90 to 104 mm SVL (97 mm). Tail length is 180% to 200% of SVL. Hatchling length is unknown.

OCCURRENCE: *E. mokolahi* is widespread in the Tongan Islands extending from the Vavaʻu group in the north to ʻEua in the south, and apparently absent from the northern Tongan outliers (Niuafoʻou, Niuatoputapu).

HABITAT CHOICE: This treeskink is semiarboreal and commonly encountered foraging on the ground in and among gardens and shrubby and weedy fencerow. It escapes by climbing the nearest tree.

REPRODUCTION: Reproductive data are lacking, although females probably lay two-egg clutches.

Emoia mokosariniveikau   **Vanualevu Slender Treeskink**
Plate 21

APPEARANCE: Midsize, slender lizard with moderately long limbs and long, tapering tail. Its head is ovate in dorsal outline and distinct from neck. Dorsal ground color is coppery brown, overlain by turquoise on neck and trunk; sides of neck and trunk have series of dark-brown bars. Head is uniform bright, coppery brown above and turquoise extending onto jowls. Ventrally chin and anterior throat are whitish, becoming light orangish yellow from midthroat onto tail. Neck, trunk, and tail have moderate-sized, smooth, imbricate scales; scales are nearly equal in size around body. Scales on head are smooth, enlarged, abutting plates; supranasal scales are present, and interparietal scale is modest, elongate pentagonal shape with parietal eye; anterior loreal is longer than high, and there are usually eight supralabials with sixth largest and beneath eye. First pair of middorsal body scales behind parietals is enlarged as nuchal plates. Middorsal scales are in 61 rows from nape to base of tail, 30 scale rows encircle body in middle of trunk, and fourth toe (hindfoot) bears 48 to 49 smooth lamellae on its underside.

SIZE: An adult female was 55 mm SVL; no males are known. Tail length is 165% of SVL.

OCCURRENCE: *E. mokosariniveikau* is known from a single forested site in central Vanua Levu.

HABITAT CHOICE: This skink was found in understory trees along the forested edge of a recently prepared garden site.

REPRODUCTION: No data are available.

Emoia nigra   **South Pacific Black Skink**
Plate 22

APPEARANCE: Midsize to large, robust lizard with long limbs and long, tapering tail. Its head is ovate in dorsal outline and distinct from neck. Dorsal ground color is uniform shiny black. Head matches body coloration. Venter from chin onto trunk is uniform dark brown. Neck, trunk, and tail have moderate-sized, smooth, imbricate scales; scales are nearly equal in size around body. Scales on head are smooth, enlarged, abutting plates; supranasal scales are present, and interparietal scale is small, triangular with parietal eye; anterior loreal is higher than long, and there are usually eight supralabials with sixth largest and beneath eye. First pair of middorsal body scales behind parietals is enlarged as nuchal plates. Middorsal scales are in 61 to 72 rows from nape to base of tail, 33 to 40 scale rows encircle body in middle of trunk, and fourth toe (hindfoot) bears 32 to 39 broad, smooth lamellae on its underside.

SIZE: Males average slightly larger than females. Adult females range from 85 to 114 mm SVL; males from 89 to 121 mm SVL. Tail length is 150% to 200% of SVL. Hatchlings range from 34 to 41 mm SVL.

OCCURRENCE: The South Pacific Black Skink is found from Vanuatu eastward to Tonga and Samoa and northward to Rotuma and Wallis and Futuna.

HABITAT CHOICE: *E. nigra* lives broadly from suburban vacant lots and gardens to forest edges and open-canopied forest. It is primarily a terrestrial forager, although it climbs low in trees.

REPRODUCTION: Females lay clutches of two to four eggs (usually two) in rotten plant debris and commonly deposit eggs in rotting logs and stumps. Incubation is approximately 65 days. Females were gravid year-round in American Samoa. This aseasonal reproduction is likely for other island groups also.

MISCELLANEA: *E. nigra*, aside from eating a variety of invertebrates, is an active predator of smaller skinks such as *E. cyanura* and *E. impar*.

Emoia nigromarginata   **Vanuatu Coppery Vineskink**

Plate 19

APPEARANCE: Midsize, medium-bodied lizard with moderately long limbs and long, tapering tail. Its head is ovate in dorsal outline and distinct from neck. Dorsal ground color is light coppery tan to coppery olive with pair of diffuse, greenish-ivory dorsolateral stripes that, on neck and anterior trunk, extend bars ventrally, thus producing alternating pattern of light and dark; broad copper-colored stripe from behind eye to hindlimb, trunk is greenish ventrolaterally from jowls to hindlimbs. Top of head matches dorsal body color, usually without dark marks. Venter is immaculate, ivory white from chin to anterior throat, yellowish ivory from anterior throat to anterior chest, and becoming yellow to greenish yellow from rear of chest onto base of tail. Neck, trunk, and tail have moderate-sized, smooth, imbricate scales; scales are nearly equal in size around body. Scales on head are smooth, enlarged abutting plates; supranasal scales are present, and interparietal scale is modest sized, elongate pentagonal shape with parietal eye; anterior loreal is longer than high, and there are usually eight supralabials with sixth largest and beneath eye. First pair of middorsal body scales behind parietals is enlarged as nuchal plates. Middorsal scales are in 56 to 61 rows from nape to base of tail, 30 to 33 scale rows encircle body in middle of trunk, and fourth toe (hindfoot) bears 35 to 42 smooth lamellae on its underside.

SIZE: Males average distinctly larger than females. Adult females range from 55 to 67 mm SVL (mean, 67 mm); males from 52 to 76 mm SVL (67 mm). Tail length is 160% to 190% of SVL. Hatchling size has not been reported.

OCCURRENCE: *E. nigromarginata* is a Vanuatuan endemic and lives on the larger islands—that is, Pentecost, Espiritu Santo, Malekula, and Efate of the northern island group.

HABITAT CHOICE: This treeskink appears most common in regenerating forest habitats and forest edges, occupying lower branches of trees and shrubby tangles.

REPRODUCTION: Gravid females carry two to three eggs. Other aspects of this skink's reproductive biology are unknown.

Emoia oriva   **Rotuma Barred Treeskink**

Plate 20

APPEARANCE: Midsize, medium-bodied skink with moderately long limbs and long, subcylindrical tail tapering to pointed tip. Head, body, limbs, and tail are coppery-brown ground color with series of transversely arranged dark-brown markings and numerous lime-green ticks from neck to base of tail; dorsally limbs are mottled with dark-brown and small, greenish-ivory spots; broad, dark-brown stripe extends from behind eye to anterior trunk and is usually continuous from the ear-opening to shoulder, where it breaks into lateral series of three or four rectangular spots. Venter is immaculate, shiny yellow on chin and anterior throat, grading into turquoise venter; undersides of feet commonly are bright golden yellow. Neck, trunk, and tail have moderate-sized, smooth, imbricate scales; scales are nearly equal in size around body. Scales on head are smooth, enlarged, abutting plates; supranasal scales are present, and interparietal scale is elongate pentagonal shape with parietal eye; anterior loreal is longer than high, and there are usually eight supralabials with sixth largest and beneath eye. First pair of middorsal body scales behind parietals is enlarged as nuchal plates. Middorsal scales are in 58 to 77 rows from nape to base of tail, 33 to 38 scale rows encircle body in middle of trunk, and fourth toe (hindfoot) bears 50 to 65 smooth lamellae on its underside.

SIZE: Females and males are equal in size. Adult females range from 75 to 85 mm SVL (mean, 82 mm); males from 73 to 90 mm SVL (84 mm). Tail length is 160% to 200% of SVL. Hatchling size is unknown.

OCCURRENCE: *E. oriva* is an endemic inhabitant of the small Rotuma islands group.

HABITAT CHOICE: This treeskink frequents forest edges and the scattered trees of gardens and villages. It typically chooses trees with low vegetation at their base; observations suggest that it sleeps within the vegetation mass at the bottom of tree trunks.

REPRODUCTION: Evidence is limited, but this skink appears to produce consistently two-egg clutches. Seasonality data are not available, although it is likely that females lay eggs year-round owing to the constancy of Rotuman climate.

Emoia parkeri   **Fiji Copper-headed Skink**

Plate 23

APPEARANCE: Midsize, slender lizard with moderately long limbs and long, tapering tail. Its head is elongate ovate in dorsal outline with pointed snout and distinct from neck. Dorsal ground color is grayish olive to light coppery brown, immaculate middorsally and bordered parasagittally by dark-brown mottling from midneck to sacrum, dark-brown stripe high on side of trunk from side of head to hindlimb, with midlateral light stripe bordered below by weak to heavy dark-brown mottling. Top of head matches body ground color, usually immaculate; dark-brown stripe from midloreal area through eye and continuous with lateral body stripe. Ventrally, chin and throat are coppery ivory becoming light yellowish green posteriorly to base of tail. Neck, trunk, and tail have moderate-sized, smooth, imbricate scales; scales are nearly equal in size around body. Scales on head are smooth, enlarged, abutting plates; supranasal scales are present; and interparietal scale is small, lanceolate, with parietal eye. Anterior loreal is longer than high, and there are usually eight supralabials with sixth largest and beneath eye. First pair of middorsal body scales behind parietals is enlarged as nuchal plates. Middorsal scales are in 54 to 60 rows from nape to base of tail, 28 to 33 scale rows encircle body in middle of trunk, and fourth toe (hindfoot) bears 31 to 41 broad, smooth lamellae on its underside.

SIZE: Females and males are equal in size. Adult females range from 43 to 52 mm SVL (mean, 49 mm); males from 47 to 50 mm SVL (49 mm). Tail length is about 150% of SVL. Hatchlings range from 23 to 26 mm SVL.

OCCURRENCE: *E. parkeri* is a Fijian endemic found on the core islands from Taveuni and Vanua Levu south to Kadavu.

HABITAT CHOICE: *E. parkeri* is a semiarboreal forest lizard, living in a variety of forest types from dry coastal ones to moist montane evergreen forest. It is seen regularly along the forest edges or in windfalls within the forest.

REPRODUCTION: Females produce two-egg clutches. The eggs are laid in soil and litter around the base of trees.

Emoia ponapea   Pohnpei Skink

Plate 23

APPEARANCE: Midsize, slender lizard with moderately long limbs and long, tapering tail. Its head is ovate in dorsal outline and distinct from neck. Dorsal ground color is medium brown with coppery overtones with scattering of small, dark-brown blotches; blotches concentrated into irregular-edged and discontinuous dorsolateral stripe from neck to hindlimbs; sides of trunk slightly darker than dorsum and with light-bluish spots; dark-brown spot behind forelimb insertion. Head matches dorsal trunk color, immaculate to lightly blotched; dark-brown lateral stripe extends from loreal area to above ear. Venter is immaculate, ivory white from chin onto tail. Neck, trunk, and tail have moderate-sized, smooth, imbricate scales; scales are nearly equal in size around body. Scales on head are smooth, enlarged, abutting plates; supranasal scales are present, and interparietal scale is moderate, diamond shaped with parietal eye; anterior loreal is higher than long, and there are usually eight supralabials with sixth largest and beneath eye. First pair of middorsal body scales behind parietals is enlarged as nuchal plates. Middorsal scales are in 52 to 62 rows from nape to base of tail, 29 to 33 scale rows encircle body in middle of trunk, and fourth toe (hindfoot) bears 38 to 45 broad, smooth lamellae on its underside.

SIZE: Females and males are equal in size. Adult females range from 44 to 50 mm SVL; males from 43 to 51 mm SVL. Tail length is about 120% to 130% of SVL.

OCCURRENCE: This small skink lives only on Pohnpei.

HABITAT CHOICE: The Pohnpei Skink is a terrestrial species of the forest of the lowlands and mountains of Pohnpei.

REPRODUCTION: Gravid females bear two eggs. Other details of this skink's reproduction are unknown.

Emoia rufilabialis   **Red-lipped Striped Skink**

Plate 18

APPEARANCE: Midsize, robust lizard with long limbs and long, tapering tail. Its head is ovate in dorsal outline and distinct from neck. Coloration is dichromatic. All individuals share ground color of brown to coppery olive dorsally, becoming dark coppery green on sides; juveniles and subadult females have a broad, creamy-yellow middorsal stripe from tip of snout onto tail and narrower cream dorsolateral stripes from above eyes to base of tail; stripes except for middorsal stripe on head and neck disappear in most males and older females; most adult females retain faded middorsal stripe on trunk but no dorsolateral stripes. Head matches body coloration except for orangish-red lower lips. Tail of both sexes is brassy green above and pale green below. Ventrally, chin and throat range from dusky ivory to pale turquoise, merging into orange to salmon chest and belly. Neck, trunk, and tail have moderate-sized, smooth, imbricate scales; scales are nearly equal in size around body. Scales on head are smooth, enlarged, abutting plates; supranasal scales are present, interparietal scale is absent, parietal eye in parietal scale; anterior loreal is narrow, distinctly higher than long, and there are seven or eight supralabials with fifth or sixth largest and beneath eye. First pair of middorsal body scales behind parietals is enlarged as nuchal plates. Middorsal scales are in 62 to 73 rows from nape to base of tail, 38 to 42 scale rows encircle body in middle of trunk, and fourth toe (hindfoot) bears 68 to 80 narrow, smooth lamellae on its underside.

SIZE: Males average larger than females. Adult females range from 46 to 58 mm SVL (mean, 51 mm SVL); males from 58 to 65 mm SVL (60 mm). Tail length is about 180% of total length. Hatchling size is unknown.

OCCURRENCE: This skink resides only on Santa Cruz Island, Solomon Islands.

HABITAT CHOICE: *E. rufilabialis* lives in forests with semi-open canopies, overgrown coconut plantations, and early-stage secondary-growth garden plots. It forages on the ground and low in bushes, logs, and other elevated ground litter. It is an active forager and seemingly always in motion, taking only brief pauses.

REPRODUCTION: Reproductive data are not available. Females probably lay two-egg clutches in shallow nest burrows beneath the forest-floor litter.

Emoia samoensis   **Pacific Robust Treeskink**

Plate 20

APPEARANCE: Midsize to large, robust skink with strong, medium-length limbs and long, subcylindrical tail tapering to pointed tip. Head, body, limbs, and tail are coppery-brown ground color with series of narrow transverse dark-brown markings from neck to base of tail. Dorsally limbs are mottled with dark-brown and small, greenish-ivory spots; trunk occasionally has greenish-ivory spotting. Venter is unicolor light greenish yellow. Head is typically unicolor above with dark-brown blotch behind eye on temporal area. Neck, trunk, and tail have large, smooth, imbricate scales; scales are nearly equal in size around body. Scales on head are smooth, enlarged, abutting plates; supranasal scales are present, and interparietal scale is small and triangular with parietal eye; anterior loreal is usually longer than high and usually eight supralabials with sixth largest and beneath eye. First pair of middorsal body scales behind parietals is enlarged as nuchal plates. Middorsal scales are in 59 to 67 rows from nape to base of tail, 30 to 34 scale rows encircle body in middle of trunk, and fourth toe (hindfoot) bears 41 to 51 lamellae on its underside.

SIZE: Females and males are equal in size, although males average slightly larger. Adult females range from 84 to 109 mm SVL (mean, 97 mm); males from 87 to 112 mm SVL (100 mm). Tail length is 160% to 200% of SVL. Hatchling size is presently unknown.

OCCURRENCE: *E. samoensis* is an inhabitant of Samoa. Recent evidence suggests that it is also present in northern Fiji.

HABITAT CHOICE: This treeskink forages on the ground and in the trees. It is mainly a resident of forests, both internally and on their edges.

REPRODUCTION: Females deposit clutches of four to seven eggs (usually five). Eggs are presumably deposited shallowly in the ground, in rotting logs and stumps; however, *E. samoensis* egg deposition sites have not yet been located. Females were gravid year-round in American Samoa, and this aseasonal reproduction is likely for other island groups as well.

Emoia sanfordi   **Toupeed Treeskink**

Plate 19

APPEARANCE: Midsize to large, robust skink with strong, medium-length limbs and very long, subcylindrical tail tapering to pointed tip. Dorsal ground color is usually bright, dark green from snout to tail tip. Head typically bears black patch, particularly on crown of head, although black marking can occupy most of dorsal surface of head; black marking can be more extensive and cover the entire body in mottled pattern, and, less commonly, background is muted gray, either unicolor or with small green spots. Venter is unicolor light greenish yellow. Neck, trunk, and tail have large smooth, imbricate scales; scales are subequal in size around body. Scales on head are smooth, enlarged, abutting plates; supranasal scales are present and interparietal scale small with parietal eye; anterior loreal is usually longer than high and usually there are eight supralabials, sixth one largest and beneath eye. First pair of middorsal body scales behind parietals is enlarged as nuchal plates. Middorsal scales are in 56 to 61 rows from nape to base of tail, 28 to 32 scale rows encircle body in middle of trunk, and fourth toe (hind-foot) bears 57 to 72 lamellae on its underside.

SIZE: Females and males are equal in size. Adult females range from 97 to 111 mm SVL (mean, 103 mm); males from 92 to 114 mm SVL (102 mm). Tail length is 180% to 210% of SVL. Hatchling size is 29 mm SVL.

OCCURRENCE: *E. sanfordi* is a Vanuatu endemic and living on the northern islands from Efate to the Torres Islands. The records of this species in the southern islands appear to be misidentifications of the endemic *E. samoensis* species group lizards living there.

HABITAT CHOICE: This species is strongly arboreal and seen only rarely on the ground. It occurs on isolated palms and trees along the coast through overgrown coconut grooves and scrubby re-growth forest to primary forest. In the latter, individuals of *E. sanfordi* are reported to be mainly gray morphotypes.

REPRODUCTION: Eggs are presumably laid on the ground beneath plant litter near tree bases. Clutch size ranges from three to seven eggs. Whether this species has a continuous or seasonal reproductive activity is unknown.

Emoia slevini   **Mariana Skink**

Plate 20

APPEARANCE: Midsize, robust skink with long limbs and long, tapering tail. Dorsal ground color is iridescent brown with back lighter brown than sides of trunk; usually row of dorsolateral white spots extends from neck onto base of tail. Venter from chin onto trunk is uniform, often orangish tan in adults. Neck, trunk, and tail have moderate-sized, smooth, imbricate scales; scales are nearly equal in size around body. Scales on head are smooth, enlarged, abutting plates; supranasal scales are present, and frontointerparietal scale is small, triangular, with parietal eye; anterior loreal is higher than long, and there are usually eight supralabials with sixth largest and beneath eye. First pair of middorsal body scales behind parietals is enlarged as nuchal plates. Middorsal scales are in 62 to 74 rows from nape to base of tail, 34 to 38 scale rows encircle body in middle of trunk, and fourth toe (hindfoot) bears 30 to 37 broad, smooth lamellae on its underside.

SIZE: Males average larger than females. Adult females range from 63 to 75 mm SVL; males from 69 to 84 mm SVL. Males can attain a maximum total length of ~180 mm, tail length is 55% to 60% of total length. Hatchling size may be about 32 mm SVL.

OCCURRENCE: *E. slevini* is a Mariana endemic, residing throughout the archipelago from Ascunion to Cocos Island, just south of Guam, although its presence has not been confirmed for all Mariana Islands. It is now extinct on Guam; the last record there is from 1945.

HABITAT CHOICE: This skink is a terrestrial resident of forest and forest-edge habitats.

REPRODUCTION: Gravid females bear two eggs; otherwise its reproductive biology is unknown. Nesting sites are assumed to be beneath forest-floor debris.

MISCELLANEA: *E. slevini* is a member of the *Emoia atrocostata* group. While formerly widespread, it appears to be declining or has disappeared recently from islands with populations of the invasive *Carlia ailanpalai.*

Emoia taumakoensis   Taumako Skink

Plate 18

APPEARANCE: Midsize, robust lizard with long limbs and long, tapering tail. Its head is ovate in dorsal outline and distinct from neck. Coloration is dichromatic. Males are uniformly medium to dark brown from head to tail, usually with coppery tint; juveniles and subadult females share ground color and have golden-yellow middorsal stripe from tip of snout to midneck and dorsolateral stripes from snout to base of tail; dorsolateral stripes faded or merge with light-colored sides of trunk in older females. Tails of both sexes are brassy green above and pale green below. Venter is cream to light yellow from chin to pubis. Neck, trunk, and tail have moderate-sized, smooth, imbricate scales; scales are nearly equal in size around body. Scales on head are smooth, enlarged, abutting plates; supranasal scales are present, interparietal scale is absent, parietal eye in frontoparietal; anterior loreal is higher than long, and there are typically seven supralabials with fifth largest and beneath eye. First pair of middorsal body scales behind parietals is enlarged as nuchal plates. Middorsal scales are in 58 to 61 rows from nape to base of tail, 32 to 35 scale rows encircle body in middle of trunk, and fourth toe (hindfoot) bears 76 to 85 narrow, smooth lamellae on its underside.

SIZE: Males average larger than females. Adult females range from 44 to 55 mm SVL (mean, 49 mm SVL); males from 50 to 59 mm SVL (54 mm). Tail length is 150% to 190% of total length. Hatchling size is unknown.

OCCURRENCE: This skink resides only in the Duff Islands, Solomon Islands.

HABITAT CHOICE: *E. taumakoensis* lives in semi-open canopy forest, forest edge, and partially cleared sites. Although principally a forest-floor resident, it uses elevated locations such shrubs and tree buttresses for basking.

REPRODUCTION: Reproductive data are not available; females probably lay two-egg clutches.

Emoia tongana   Polynesia Slender Treeskink

Plate 21

APPEARANCE: Midsize, slender lizard with moderately long limbs and long, tapering tail. Its head is ovate in dorsal outline and distinct from neck. Dorsal ground color is coppery olive to coppery tan, speckled with small, dark-brown marks especially on rear of head to above shoulder, elsewhere less prominent. Head matches body coloration; scales often narrowly dark edged. Venter from chin onto trunk is uniform yellowish green; soles of hindfeet are dusky green or yellow. Neck, trunk, and tail have moderate-sized, smooth, imbricate scales; scales are nearly equal in size around body. Scales on head are smooth, enlarged, abutting plates; supranasal scales are present, and interparietal scale is small and filled by parietal eye; anterior loreal is longer than high, and there are usually eight supralabials with sixth largest and beneath eye. First pair of middorsal body scales behind parietals is enlarged as nuchal plates. Middorsal scales are in 62 to 71 rows from nape to base of tail, 31 to 36 scale rows encircle body in middle of trunk, and fourth toe (hindfoot) bears 41 to 55 smooth lamellae on its underside.

SIZE: Females are smaller than males. Adult females range from 53 to 71 mm SVL (mean, ~61 mm); males from 55 to 75 mm SVL (~64 mm). Tail length is 160% to 200% of SVL. Hatchling size is presently unknown.

OCCURRENCE: *E. tongana* is a south-central Pacific species with populations on Futuna and Savai'i and Upolo (Samoa) southward to the Ha'apai island group of Tonga.

HABITAT CHOICE: This slender treeskink is arboreal and seen most frequently on heavily treed fencerows, forest edges, and within open-canopied and scrubby forest.

REPRODUCTION: Clutch size has not been reported for this species; presumably it is the typical two eggs per clutch. Other details of its reproduction are also unknown.

Emoia trossula   **Fiji Barred Treeskink**

Plate 20

APPEARANCE: Midsize to large, robust skink with strong, moderately long limbs and long, subcylindrical tail tapering to pointed tip. Head, body, limbs, and tail are coppery-brown ground color with variable pattern of dorsal and lateral markings on neck and trunk ranging from near unicolor with scattering of small, dark blotches and small (single scale) pale-green spots to, more commonly, series of transversely arranged dark-brown markings and numerous lime-green ticks from neck to base of tail; dorsally limbs are mottled with dark-brown and small, greenish-ivory spots; broad, dark-brown stripe extends from behind eye to above ear-opening. Venter is immaculate, usually ivory white on chin and becoming greenish yellow on anterior throat and strikingly yellowish green on abdomen; undersides of fore- and hind-feet commonly are yellow, dull to bright. Neck, trunk, and tail have moderate-sized, smooth, imbricate scales; scales are nearly equal in size around body. Scales on head are smooth, enlarged, abutting plates; supranasal scales are present, and interparietal scale is small elongate pentagonal with parietal eye; anterior loreal is longer than high, and there are usually eight supralabials with sixth largest and beneath eye. First pair of middorsal body scales behind parietals is enlarged as nuchal plates. Middorsal scales are in 59 to 69 rows from nape to base of tail, 31 to 38 scale rows encircle body in middle of trunk, and fourth toe (hindfoot) bears 41 to 60 lamellae on its underside.

SIZE: Females and males are equal in size. Adult females range from 81 to 101 mm SVL (mean, 94 mm); males from 80 to 107 mm SVL (97 mm). Tail length is 180% to 200% of SVL. Hatchling size has not been reported.

OCCURRENCE: When described in the late 1980s, *E. trossula* was thought to be widespread throughout the south-central Pacific except for Samoa. That concept is no longer valid. *E. trossula* is now considered to have a much smaller distribution, limited to northern Fiji—principally but not exclusively to Viti Levu, Vanua Levu, Taveuni, and adjacent islands and islets.

HABITAT CHOICE: *E. trossula* is an arboreal lizard of forest edges, open-canopied forest, treed fencerows, and coconut plantations.

REPRODUCTION: Like other robust bodied *E. samoensis* group species, *E. trossula* females probably produce clutches of usually two and three eggs (maximum five eggs) and deposit them in terrestrial nests beneath vegetation debris near or at the base of trees.

Emoia tuitarere   **Rarotonga Treeskink**

Plate 20

APPEARANCE: Midsize, medium-bodied skink with moderately long limbs and long, subcylindrical tail tapering to pointed tip. Head, body, limbs, and tail are coppery-brown ground color with series of transversely arranged dark-brown markings and numerous lime-green ticks from neck to base of tail; dorsally limbs are mottled with dark-brown and small, greenish-ivory spots; broad, dark-brown stripe extends from eye to axilla. Venter is immaculate, yellow on chin and anterior throat; thereafter onto to tail it is light greenish yellow with dense, dark-brown speckling. Neck, trunk, and tail have moderate-sized, smooth, imbricate scales; scales are nearly equal in size around body. Scales on head are smooth, enlarged, abutting plates; supranasal scales are present, and interparietal scale is moderate-sized with parietal eye; anterior loreal is longer than high, and there are usually eight supralabials with sixth largest and beneath eye. First pair of middorsal body scales behind parietals is enlarged as nuchal plates. Middorsal scales are in 62 to 68 rows from nape to base of tail, 32 to 34 scale rows encircle body in middle of trunk, and fourth toe (hindfoot) bears 46 to 53 broad, smooth lamellae on its underside.

SIZE: Females and males are equal in size. Adult females range from 72 to 93 mm SVL (mean, 81 mm); males from 68 to 93 mm SVL (84 mm). Tail length is ~200% of SVL. Hatchling size is unknown.

OCCURRENCE: *E. tuitarere* occurs on a single island (Rarotonga) in the Cook Islands. It is uncertain whether this treeskink is truly an endemic species or an alien species recently introduced into Rarotonga from an unknown source island. The earliest record of *E. tuitarere* in Rarotonga is from the 1980s.

HABITAT CHOICE: These treeskinks inhabit forest edges, wood lots, broken-canopied scrub, and secondary-growth forest. They are largely arboreal and rarely encountered on the ground, but are often found low in branches of understory trees and shrubs.

REPRODUCTION: Reproductive data are unavailable; presumably most females produce two-egg clutches.

Eugongylus albofasciolatus   **Barred Recluse Skink**

Plate 24

APPEARANCE: Large, elongate, and robust-bodied lizard with short, well-developed limbs and heavy, thick tail. Its head is large, triangular, and posteriorly barely wider than neck. Dorsal ground color is medium brown to chocolate brown with numerous light (white to yellow) transverse bars on neck and trunk; bars have speckled appearance and are widest dorsally and narrow ventrally, also broader anteriorly and narrower posterior. Head is nearly uniform dorsally and laterally; upper and lower lips are lighter with three to four dark-brown streaks that curve downward onto throat. Venter is uniform creamy white with dark labial streaks onto unicolor throat, venter often with an orangish tint from base of neck onto tail. Neck, trunk, and tail have large smooth, imbricate scales; scales are subequal in size around body. Scales on head are smooth, enlarged, abutting plates; supranasal scales are present; interparietal scale large, elongate pentagonal, with parietal eye; there are eight supralabials, fifth one largest and beneath eye although separated from the orbit by small subocular scales, seventh and eighth supralabials horizontally divided. First pair of middorsal body scales is enlarged as nuchal plates. Middorsal scales are in 76 to 84 rows from nape to base of tail, 35 to 38 scales encircle body in middle of trunk, and fourth toe (hindfoot) bears 16 to 19 lamellae on its underside.

SIZE: Adult males are distinctly larger than females. Adult females range from 103 to 130 mm SVL (mean, ~118 mm); males from 138 to 150 mm SVL (~145 mm). Tail length is approximately 120% to 160% of SVL. Hatchling size has not been reported.

OCCURRENCE: *E. albofasciolatus* has a divided distribution. The southern distribution is the north coast New Guinea and the Solomon Islands eastward into the Santa Cruz group. The northern distribution extends spottily from Chuuk of the Caroline Islands to Arno Atoll in the Marshall Islands.

HABITAT CHOICE: This surface-dweller is a resident of garden litter and coconut husk piles in villages, and is also found amid the leaf litter and beneath logs in both disturbed and little-disturbed forest. It is uncommonly seen and active mainly at dusk and night.

REPRODUCTION: Other *Eugongylus* species lay two to four eggs; a similar clutch size is likely for this species as well.

MISCELLANEA: *E. albofasciolatus* appears to be a catholic carnivore, capturing and consuming a large variety of invertebrate and small vertebrate prey.

Although individuals from the northern and southern populations appear similar, there are a few morphological features hinting at the presence of two species. The above appearance description derives from individuals of the Caroline population.

Eugongylus nsp   **Palau Recluse Skink**

Not illustrated

APPEARANCE: Large, elongate and robust-bodied lizard with short, well-developed limbs and heavy thick tail. Its head is large, triangular, and posteriorly barely wider than neck. Dorsal ground color is light to medium brown anteriorly, becoming orangish brown ventrolaterally on trunk; neck and trunk bear numerous small black tics and yellowish-white spots; posterior third of trunk and much of tail are dark brown to black and obscure light spotting. Top and side of head is unicolor light to medium brown; upper and lower lips are lighter, with three dark-brown streaks that curve downward onto throat. Venter from chin and throat is white, often with pinkish tint; chest and belly are salmon to dull orange, and tail is white to faint salmon. Neck, trunk, and tail have large, smooth, imbricate scales; scales are subequal in size around body. Scales on head are smooth, enlarged, abutting plates; supranasal scales are present and interparietal scale large, triangular, with parietal eye; there are eight supralabials, sixth one largest and beneath eye and separated from orbit by small subocular scales. First pair of middorsal body scales is enlarged as nuchal plates. Middorsal

Scales on head are smooth, enlarged, abutting plates; supranasal scales are absent and interparietal scale medium-sized and pentagonal with parietal eye; there are seven supralabials, fifth one beneath eye. First pair of middorsal body scales is enlarged as nuchal plates. Middorsal scales are in 51 to 60 rows from nape to base of tail, 28 to 32 scale rows encircle body in middle of trunk, and fourth toe (hindfoot) bears 22 to 25 lamellae on its underside.

SIZE: Adult females and males are equal in size. Adult females range from 38 to 53 mm SVL; males from 40 to 53 mm SVL. Tail length is 130% to 150% of SVL.

OCCURRENCE: With the exception of Efate, *C. atropunctatus* is confined to the southern islands of Erromango, Aniwa, Tanna, Anatom, and Futuna. This skink is widespread on New Caledonia and the Loyalty Islands. This commonness in the latter two areas suggests that its presence on Vanuatu results from human transport rather than natural dispersal, although its widespread occurrence in the islands argue for an older, natural dispersal event, perhaps during the last glacial period when sea levels were lower.

HABITAT CHOICE: This skink is broadly tolerant of habitats that include leaf litter; hence it regularly lives in dry to moist forest, gardens, and horticultural landscapes.

REPRODUCTION: Reproductive data are limited to clutches two to three eggs.

MISCELLANEA: *C. atropunctatus* is a forest-floor resident seemingly always on the move in and out of the leaf litter in search of invertebrate prey.

Lamprolepis smaragdina   **Emerald Treeskink**

Plate 19

APPEARANCE: Midsize to large, medium-bodied lizard with moderately long limbs and long, tapering tail. Its head is ovate in dorsal outline and distinct from neck. Dorsal ground color is variable within and between populations; basic ground color is shiny dark green, typically with each scale edged in black; commonly the posterior half of trunk and hindlimbs are light coppery brown; other, less-frequent ground colors are dull olive to dull yellowish brown. Head matches body coloration, usually with black-edged scale. Venter is also variable, ranging from pale green to yellow; in some base of throat and chest is turquoise;

underside of tail usually matches the abdomen; heels and soles of hind-feet are dark yellow. Scales are smooth, enlarged as abutting plates on head, overlapping on remainder of body, and moderate-sized and subequal dorsally to ventrally on neck and trunk. On head, supranasals are absent, interparietal is of modest size and roughly diamond-shaped, anterior loreal is much longer than high, and the enlarged sixth supralabial is beneath eye. Middorsal scales are in 44 to 84 rows from nape to base of tail, 22 to 24 scale rows encircle body in middle of trunk, and fourth toe (hindfoot) bears 24 to 34 broad, smooth lamellae on its underside.

SIZE: Females average slightly larger than males. Adult females range from 82 to 107 mm SVL; males from 80 to 100 mm SVL. Tail length is 150% to 180% of SVL. Hatchlings range from 31 to 42 mm SVL.

OCCURRENCE: The Emerald Treeskink lives throughout the western Pacific and Pacific Rim islands.

HABITAT CHOICE: This treeskink is seen regularly on the trunks of trees one or two meters above the ground in yards, gardens, and other disturbed habitats. The trees selected are typically isolated and surrounded by low vegetation with a nearby cover of shrubs or denser ground vegetation.

REPRODUCTION: *L. smaragdina* females bear two eggs; these appear to be deposited in a variety of microhabitats, from high in palm-leaf axils to terrestrial nests beneath logs or rocks. Evidence from the eastern Papua New Guinea population indicates year-round reproductive activity and the likelihood that females deposit more than one clutch each year. Philippine populations also reproduce year-round.

MISCELLANEA: The diet of the Emerald Treeskink is broad and includes invertebrates, small vertebrates, and even fruit. It forages throughout the day from treetop to the ground. To escape when on a tree trunk it moves to the opposite side of the tree and upward; it will also occasionally descend to the ground and dash into thick vegetative cover.

Early in the 20th century, a German herpetologist partitioned *L. smaragdina* into eight subspecies. This concept receive early acceptance, but late-20th-century observations recognized that the ranges of the proposed subspecies did not match the distribution of the various color morphs. Hence, the use of subspecific names has nearly disappeared. Unfortunately, no one has yet carefully examined the morphology of *L. smaragdina* to determine whether a single very variable species or multiple species exist. The latter situation seems the most likely.

Lampropholis delicata   **Garden Skink**

Plate 26

APPEARANCE: Small to midsize, cylindrical-bodied lizard with short, robust limbs and thick, cylindrical tail. Its head is small, triangular, and posteriorly barely wider than neck. Dorsal ground color is light to medium brown or orangish brown with darker-brown mottling dorso-laterally and laterally; dark mottling can form ragged-edged dorsolateral stripe in some individuals. Head is less mottled than neck and trunk, often more brightly colored, and lips are dark and light barred. Venter from chin onto trunk is uniform cream to light tan, posterior trunk and tail commonly orangish. Neck, trunk, and tail have large smooth, imbricate scales; scales are subequal sized around body. Scales on head are smooth, enlarged, abutting plates; supranasal scales are absent and interparietal scale large, triangular, with parietal eye; there are six supralabials, fourth one largest and beneath eye although separated from the orbit by small subocular scales. First three or four rows of middorsal body scales are enlarged as nuchal plates. Middorsal scales are in 55 to 62 rows from nape to base of tail, 23 to 26 scales encircle body in middle of trunk, and fourth toe (hindfoot) bears 12 to 17 lamellae on its underside.

SIZE: Females and males are equal in size. Adult females range from 33 to 46 mm SVL (mean 39 mm); males from 31 to 42 mm SVL (37 mm). Tail length is 150% to 180% of SVL. Hatchlings range about 15 to 17 mm SVL.

OCCURRENCE: This Australian skink arrived in Oʻahu about 1917, presumably as a cargo stowaway. It is now widespread in a broad array of disturbed habitats from vacant city lots to abandon farmlands and into moist native forest, occurring on the islands of Hawaiʻi, Kauaʻi, Lanaʻi, Maui, Molokaʻi, and Oʻahu. In Australia, it is an east coast skink ranging from Cairns, Queensland, to Melbourne, Victoria, and into Tasmania.

HABITAT CHOICE: This alien occurs almost everywhere and becomes abundant in microhabitats with a thick groundcover of leaves and other plant detritus. Its superabundance in many sites has resulted in its label of the Plague Skink in Hawaii.

REPRODUCTION: *L. delicata* appears to reproduce year-round in the Hawaiian Islands, and it does so prolifically with a clutch size of one to seven eggs, usually three or more. Eggs are laid beneath plant debris and other items lying on the ground surface. In Australia, females are reproductively active from September to March, and males appear reproductively ready year-round.

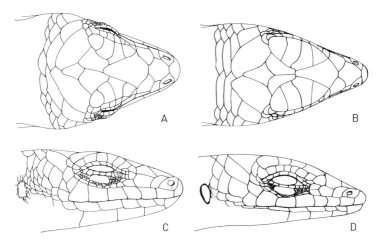

FIGURE 17. Head morphology of two Central Pacific terrestrial skinks with robust, elongate bodies and short limbs. Dorsal and lateral views of the head of *Leiolopisma alazon* (A, C) and *Lipinia noctua* (B, D).

Leiolopisma alazon   **Ono-i-Lau Ground Skink**

Plate 24, Figure 17

APPEARANCE: Midsize, elongate, cylindrical-bodied lizard with short, robust limbs and thick, cylindrical tail. Its head is small, triangular, and posteriorly barely wider than neck. Dorsal ground color is medium brownish olive with scattering of light golden beige and dark-brown spots; dorsolaterally, dense series of dark spots forms irregular dark stripe from eye to hindlimb. Dorsally, head is immaculate, rufous over snout, light-brown posterior and dark brown over eyes; lips are dark and light barred. Tail varies from creamy beige to rufous orange (salmon) above and below. Venter from chin to posterior trunk is uniform cream to light tan. Neck, trunk, and tail have large smooth, imbricate scales; scales are subequal in size around body. Scales on head are smooth, enlarged, abutting plates; supranasal scales are absent and interparietal scale is moderately large, pentagonal with parietal eye; there are seven supralabials, fifth one largest and beneath eye although separated from orbit by small subocular scales. First pair of middorsal body scales is enlarged as nuchal plates. Middorsal scales are in 71 to 77 rows from nape to base of tail, 34 to 37 scales encircle body in middle of trunk, and fourth toe (hindfoot) bears 20 to 22 lamellae on its underside.

SIZE: Males seem to average larger than females. Adult females range from 43 to 48 mm SVL; males from 51 to 60 mm SVL. Tail length is about 100% to 110 % of SVL. Hatchling size is unknown.

OCCURRENCE: This skink was discovered on a single islet in the Oni-i-Lau atoll in Fiji. It was absent from nearby islets and the main island of this cluster. Fortunately, a survey conducted in 2011 of the original island discovered that it persisted there and was also present on at least two other islets within the Ono-i-Lau cluster.

HABITAT CHOICE: A surface-dweller of forested islets that have a complete canopy, largely shrubfree understory, and a thin leaf-litter surface cover. Perhaps it forages during the day; however, nocturnal foraging seems most likely because diurnal searching revealed most individuals beneath logs on the forest floor.

REPRODUCTION: Reproductive data are limited. It is not clear whether this skink is an egg-layer or livebearer.

Mabuyidae: Multiple-keeled Skinks

Until recently, this family, as a subfamily or species group, consisted largely of the single cosmopolitan genus *Mabuya* with its numerous species. All shared a stocky, moderately short body and sturdy, medium-length limbs; the majority also shared moderately large scales with strong keeling, usually multiple keels on each dorsal and lateral scale. This characterization now applies to the family, which is becoming generically more diverse as molecular studies reveal more and more lineages within the former *Mabuya*.

Mabuyids are typically diurnal, active-foraging lizards that generally search for prey on the ground or on rocks and logs lying on the ground. Most species are forest to forest-edge species, living in the full spectrum of open woodlands to closed, canopied forests. Reproductively, they have oviparous and viviparous members.

Eutropis multicarinata   **Micronesia Multi-keeled Sunskink**
Plate 25, Figure 13
APPEARANCE: Midsize, robust lizard with moderate length limbs and medium-length, tapering tail. Its head is ovate in dorsal outline and distinct from neck. Dorsal ground color is dark medium brown and unicolor in adult males; juveniles have white dorsolateral stripe from above eye onto tail, broad, dark, rufous-brown stripe from eye to

hindlimb and white to light-blue stripe from beneath eye to forelimb, continuing as series of white spots between limbs. Juvenile coloration is gradually lost during sexual maturation, yielding the unicolor male and older females, although subadults of both sexes and young mature females retain some of the pattern, especially anteriorly. Venter from chin onto tail is immaculate, chin and throat white, and abdomen usually bluish white onto tail. Scales are imbricate, strongly multi-keeled dorsally and laterally on neck, trunk, limbs, and tail, smooth ventrolaterally and ventrally, moderate-sized and subequal dorsally to ventrally on neck and trunk; head scales are rugose to smooth, large plates. On head, supranasals are present, interparietal is moderately large, pentagonal, and usually without parietal eye; anterior loreal is roughly square; and the enlarged fifth or sixth supralabial is beneath eye. Middorsal scales are in 40 to 44 rows from nape to base of tail, 30 to 33 scale rows encircle body in middle of trunk, and fourth toe (hindfoot) bears 21 to 26 broad, smooth lamellae on its underside.

SIZE: Females and males are equal in size. Adult females range from 65 to 73 mm SVL (mean, ~69); males from 62 to 76 mm SVL (~70 mm). Tail length is 130% to 200% of SVL. Hatchlings range from about 26 to 32 mm SVL.

OCCURRENCE: *E. multicarinata* is widespread throughout the Philippine Islands. The Palauan *Eutropis* occurs widely, although not throughout the entire Palauan Archipelago.

HABITAT CHOICE: This skink is largely a forest-floor resident, seen foraging in the floor litter, on rotten trunks, and low on the trunks of trees. It lives in a variety of forests types, from scrubby secondary-growth forests to primary ones.

REPRODUCTION: Reproductive data are limited. Clutch size is usually two eggs. The eggs are laid beneath leaf litter and rotten logs.

MISCELLANEA: Crombie and Pregill, in their study of Palauan herpetofauna, reported that the pre-1980 concept of *M. multicarinata* as a single species encompassing populations from Borneo and Palau has been altered. *M. multicarinata* now consists of four taxa in the Philippines, and the status of the Borneo and the Palauan populations remains undetermined. Crombie and Pregill preferred to recognize the Palauan skink as an unidentified species, which is treated here as *E. multicarinata*, although the preceding description of its morphology is restricted to individuals from Palau because this population may represent a new species.

Sphenomorphidae: Twilight Skinks

Sphenomorphid skinks, with the exception of the Moth Skinks, occur only peripherally in the Pacific. They have their greatest diversity in Australia and New Guinea, and are less diverse yet widespread in tropical and subtropical Asia. Most species are active under low-light conditions. The diurnal species occur predominantly in well-canopied forest and forage on the surface beneath leaf and other plant litter. Other species are crepuscular or strictly nocturnal, foraging primarily in and on the surface of the ground litter. In association with their life-style, most sphenomorphids are small to medium-sized with elongate cylindrical bodies and tails; characteristically, the fore- and hindlimbs are short and widely separated. Reproduction includes both oviparity and viviparity.

Lipinia leptosoma   **Pandanus Moth Skink**

Plate 24

APPEARANCE: Small, slender lizard with medium-length, slender limbs and cylindrical tail. Head is triangular in outline with pointed snout and distinctly wider posterior than adjoining neck. Dorsum is strikingly marked; middorsal stripe extends from snout onto tail, bright white on head becoming yellow to gold on trunk and grading into lemon yellow to orange-yellow posteriorly; middorsal stripe edged from head onto tail base by moderately wide, dark-brown stripe; a broad, tan dorsolateral stripe extends from eye to base of tail, and a narrow, dark-brown stripe runs from loreal area to anterior trunk before becoming series of dark spot to hindlimb. Tail ground color is bright yellow to orange-yellow with dark dorsolateral stripe disrupted anterior and disappearing by midlength. Ventrally, chin and anterior throat are generally grayish white, becoming white with coppery overtones on neck and trunk, and yellow on tail. Neck, trunk, and tail have large smooth, imbricate scales; middorsal two rows are largest and scales progressively decrease in size to ventrolateral region, becoming subequal across venter although still large. Scales on head are smooth, enlarged, abutting plates; supranasal scales are absent and interparietal scale large, elongate pentagonal shape with parietal eye; there are seven supralabials, fifth one largest and beneath eye. First three to five pairs of middorsal body scales behind parietals are only slightly larger than subsequent middorsal body scales. Middorsal scales are in 46 to 54 rows from nape to base of tail, 22 to 26 scales encircle body in middle of trunk, and fourth toe (hindfoot) bears seven to 10 enlarged lamellae on its underside, total lamellae about 14 to 18.

SIZE: Size data are for adult females; size data are not available for males. Adult females range from 35 to 44 mm SVL. Tail length is about 150% to 160 % of SVL. Hatchling size is unknown.

OCCURRENCE: *L. leptosoma* is a Palauan endemic, living on the larger islands of Babeldaob, Ngeaur, and Oreor.

HABITAT CHOICE: This skink was originally proposed as a pandanus obligate. Although pandanus is the preferred habitat, individuals appear to survive and reproduce in palms where the pandanus forest have been decimated or even eliminated.

REPRODUCTION: Data on any aspects of this species reproduction are lacking.

Lipinia noctua   **Pacific Moth Skink**

Plate 24, Figure 17

APPEARANCE: Barely midsize, elongate, stout-bodied lizard with short, robust limbs and thick, cylindrical tail. Its head is triangular in outline with bluntly pointed snout, posteriorly slightly wider than neck. Dorsal ground color is light to medium brown with broad, light-colored middorsal stripe; stripe begins with bright white to cream parietal spot accentuated by dark-brown border, becoming beige on neck and extending onto base of tail; parasagittal and dorsolateral areas are densely spotted with dark brown; laterally, interrupted dark-brown stripe extends from in front of eye onto tail, and lower lateral area is light beige stippled with dark brown. Venter is white to bright cream from chin onto tail, lightly stippled with dark brown on chin and anterior throat, and on pubis and base of tail. Tail is lighter beige than middorsal stripe and retains trace of dark-brown lateral stripe anteriorly. Neck, trunk, and tail have large smooth, imbricate scales; scales are mainly subequal in size around body, although middorsal pair is slightly larger than adjacent scales, Scales on head are smooth, enlarged, abutting plates; supranasal scales are absent and interparietal scale large, triangular, with parietal eye; there are seven supralabials, fifth one largest and beneath eye although separated from orbit by small subocular scales. First two to three pairs of middorsal body scales are enlarged as nuchal plates. Middorsal scales are in 46 to 57 rows from nape to base of tail, 24 to 28 scales encircle body in middle of trunk, and fourth toe (hindfoot) bears 17 to 25 lamellae on its underside.

SIZE: Females average larger than males. Adult females range from 37 to 47 mm SVL (mean 43 mm); males from 38 to 43 mm SVL (40 mm). Neonates are 15 to 17 mm SVL.

OCCURRENCE: *L. noctua* occurs throughout the central and western Pacific islands.

HABITAT CHOICE: The Moth Skink is forest-floor dweller in many areas, in situations ranging from heavy horticultural plantings around building to scrubby forest and edges and open areas in less-disturbed forest. In other areas, this skink seems to be strongly arboreal and is often associated with the crowns of palms.

REPRODUCTION: *Lipinia noctua* is a livebearer, seemingly with different litter sizes in different geographic regions and perhaps associated with trunk volume of gracile and robust body forms. The gracile morph typically bears one, occasionally two neonates; the robust morph usually produces two neonates and as many as four. The length of pregnancy is unknown. Evidence from Samoa indicates year-round reproductive activity.

MISCELLANEA: Crombie and Pregill, in their study of the Palauan herpetofauna, suggest two Oceania *noctua:* a heavier bodied, brightly marked Melanesian-Polynesian form and a more gracile Micronesian form with a more subdued dorsal pattern. They also report different habitat preferences, with the latter form being strongly arboreal and former one terrestrial to semiarboreal.

Prasinohaema virens   **Green-blooded Vineskink**

Plate 21

APPEARANCE: Midsize, medium-bodied lizard with moderately long limbs and long, tapering, prehensile tail. Its head is ovate in dorsal outline and distinct from neck. Dorsal ground color is light green to olive green, often with diffuse yellow middorsal stripe. Head matches body coloration. Venter from chin to trunk is unicolor greenish yellow. Scales are grossly smooth, overlapping on neck, body, and tail; first three to five pairs of middorsal scales are enlarged as nuchal scales, subsequent middorsal pair of scales about one and a half times larger than adjacent scales, and scales gradually decrease in size to the ventrolateral surface and then are uniform across venter. Dorsal trunk scales have microrugose surface; laterally, scale surface becomes smoother and totally smooth on ventral scales. On head, supranasal is absent, interparietal is large, diamond-shaped with parietal eye; anterior loreal is longer than high, and generally nine supralabial scales with seventh one beneath eye. Middorsal scales are in 57 to 64 rows from nape to base of tail, and 30 to 36 scale rows encircle body in middle of trunk. Digits have expanded pad at base of each digit; fourth toe (hindfoot) bears 13 to 17 lamellae on pad and eight to nine on terminal portion of digit.

SIZE: Males and females are equal in size. Adult females range from 47 to 52 mm SVL (mean 49 mm); males from 46 to 52 mm SVL (50 mm). Tail length is 120% to 150% of SVL. Hatchling size is about 27 mm SVL or less.

OCCURRENCE: *P. virens* occurs from northeastern New Guinea through the Bismarck Archipelago and all of the Solomon Islands to the Santa Cruz group.

HABITAT CHOICE: This vineskink is strongly arboreal and seldom voluntarily descends to the ground. It prefers dense tangles of tree limbs and branches and vines.

REPRODUCTION: *P. virens* lays clutches of two eggs in accumulations of decomposing leaves in epiphytes and tree forks or at the base of trees. Little else is known about its reproductive behavior.

MISCELLANEA: All members of the genus *Prasinohaema* have green blood plasma, which yields a greenish or bluish tint to internal tissues including the oral mucosa. Also all members of the genus have prehensile tails with a small cluster of glandular scales at their tips.

Sphenomorphus scutatus   **Palau Ground Skink**

Plate 25

APPEARANCE: Small to midsize, cylindrical-bodied lizard with short, robust limbs and thick, cylindrical tail. Its head is small, triangular, and posteriorly barely wider than neck. Dorsal ground color is light to medium brown or orangish brown with darker-brown mottling dorsolaterally and laterally; the dark mottling can form a ragged-edged dorsolateral stripe in some individuals. Head is less mottled than neck and trunk, often more brightly colored, and lips are dark and light barred. Venter from chin onto trunk is uniform cream to light tan, posterior trunk and tail commonly orangish. Neck, trunk, and tail have large smooth, imbricate scales; scales are subequal in size around body. Scales on head are smooth, enlarged, abutting plates; supranasal scales are absent and interparietal scale large, triangular, with parietal eye; there are six supralabials, fourth one largest and beneath eye although separated from orbit by small subocular scales. First three or four rows of middorsal body scales are enlarged as nuchal plates. Middorsal scales are in 55 to 62 rows from nape to base of tail, 23 to 26 scales encircle body in middle of trunk, and fourth toe (hindfoot) bears 12 to 17 lamellae on its underside.

SIZE: Females and males are equal in size. Adult females range from 33 to 44 mm SVL (mean 39 mm); males from 31 to 42 mm SVL (37 mm). Hatchling size is unknown.

OCCURRENCE: This small skink is a Palauan endemic and occurs broadly throughout the archipelago.

HABITAT CHOICE: Although *S. scutatus* was presumably originally a forest and forest-edge inhabitant, it now occurs in human-modified habitats from vacant lots and horticultural landscape through gardens into forest.

REPRODUCTION: Reproductive data are limited to females laying clutches of two eggs. These eggs are laid on the forest floor beneath detritus.

MISCELLANEA: A larger, long-bodied, short-limbed species occurs on Babeldaob and the immediately adjacent small islands to the south. Aside from its size (adults 66–73 mm SVL), it is readily distinguished from *S. scutatus* by a middorsal series of dark spots from the nape to the base of the tail, large dark blotches laterally from ear to shoulder, and its more numerous rows of dorsal scales (78–85), midbody scales (30–35), and fourth toe lamellae (18–21). It is mentioned here without its own species account because Crombie and Pregill, in their work on the herpetofauna of the Palau Islands, only tentatively referred it to the genus *Sphenomorphus*.

Sphenomorphus solomonis   **Solomon Ground Skink**

Plate 25

APPEARANCE: Midsize, elongate robust-bodied lizard with very short, robust limbs and thick, cylindrical tail. Its head is small, triangular, and posteriorly barely wider than neck. Dorsal ground color is light to medium brown with darker-brown small irregular spots from neck onto tail. Top of head is nearly uniform ground color; upper and lower lips are white with scale edge (sutures) dark gray yielding barred pattern; often gray extends onto chin and anterior throat. Venter ground color from chin onto trunk is white to light tan, anterior throat regularly speckled with small, dark-gray spots and bars. Tail is same color as body. Neck, trunk, and tail have large smooth, imbricate scales; scales are subequal in size around body. Scales on head are smooth, enlarged, abutting plates; supranasal scales are absent and interparietal scale is large, diamond-shaped, with parietal eye; there are seven supralabials, fifth one largest and beneath eye. First three or four pairs of middorsal body scales are enlarged as nuchal plates. Middorsal scales are in 65 to 72 rows from back of head to base of tail, 26 to 31 scales encircle body in middle of trunk, and fourth toe (hindfoot) bears 14 to 16 lamellae on its underside.

SIZE: Females and males are probably equal in size. Adult females range from about 49 to 57 mm SVL (mean 54 mm); males from 50 to 57 mm SVL (53 mm). Tail length is about 110% to 135% SVL. Hatchling size is unknown. (These data apply only to *S. solomonis* from the Santa Cruz island group.)

OCCURRENCE: *S. solomonis* has a long linear distribution from the Mollucas across northern New Guinea and through the Solomon Islands into the Santa Cruz group of islands.

HABITAT CHOICE: *S. solomonis* is a terrestrial skink that lives in and just beneath the surface litter of gardens and plantations into primary forest. It appears to be active at twilight and during the night.

REPRODUCTION: Data are conflicting and may represent geographic differences in reproduction. In one report, females lay two, rarely three, eggs; in another report, they lay three to six eggs. No data are available on seasonality of reproduction or duration of incubation.

ANGUIMORPHA: GLASS LIZARDS, GALLIWASPS, MONITORS, AND RELATIVES

Anguimorph lizards live in both the Old and New World and contain eight families: the Anguidae, Anniellidae, Diploglossidae, Helodermatidae, Lanthanotidae, Shinisauridae, Varanidae, and Xenosauridae. They exhibit a diversity of lifestyles from burrowing to arboreal; a few are even aquatic. As a group, they have a worldwide distribution, but five families have low species diversity and small distributions. The Anniellidae (North American Legless Lizards), Helodermatidae (Gila Monster), Lanthanotidae (Earless Monitor), and Shinisauridae (Chinese Crocodile Lizard) have only one or two species; Xenosauridae (Knob-scaled Lizards) has six species in Mexico and northern Central America. The anguids or Glass Lizards are more diverse, with about 60 species of limbless lizards (Glass Lizards, Slow Worms) and limbed species (Alligator Lizards) living throughout North and northern Central America and Euroasia. Galliwasps or diploglossid lizards include roughly 50 species of tropical American species; all are limbed, although limbs are reduced in some species. The Varanidae are the monitors and goannas.

Diploglossidae: Galliwasps and Relatives

Diplogossids lizards are strictly a New World group and are almost exclusively tropical in distribution. They are found from southern Mexico to Ecuador and central Argentina, and in the West Indies. The family contains three genera: *Celestus* with 28 species and *Diploglossus* with 18 species, both taxa sharing a similar stout, elongate cylindrical body with short, well-developed limbs; and *Ophiodes* with four snake-like species with no forelimbs and only small rudimentary hindlimbs. Diploglossids contain both egg-layers and livebearers. In both reproductive types, the number of eggs or fetuses is strongly correlated with body size. The larger taxa can produce more than 20 offspring.

Diploglossus millepunctatus   Malpelo Galliwasp

Plate 25

APPEARANCE: Large, thick, long, cylindrical-bodied lizard with pointy head; short, robust limbs; and thick, nearly blunt tail. Dorsal ground color is dark brown to nearly black with iridescent sheen and profusion of yellowish-white to pale-yellow flecks. Venter is creamy white to brownish; darker coloration is produced by ventral scale being white centered with brown edging. Body covered with shiny, moderate-sized, overlapping scales. All scales are smooth and underlain by osteoderms; neck, trunk, and caudal scale are nearly equal in size dorsally and ventrally, laterally scales distinctly smaller, and caudal scales about one and a half times larger than trunk scales. Head bears enlarged, smooth scales from snout to nape; large interparietal scale bears small parietal eye. Middorsal scales are in about 92 to 94 rows from nape to base of tail, 54 to 57 scales encircle body in middle of trunk, and fourth toe (hindfoot) bears 17 to 18 lamellae on its underside.

SIZE: Adult females and males appear equal in size, with SVL ranging from 190 to 240 mm. Tail length is about 80% to 102% of SVL.

OCCURRENCE: *D. millepunctatus* is a Malpelo endemic.

HABITAT CHOICE: The Mapelo Galliwasp occurs throughout the island from the sea edge to the summit of the two extinct volcanic cones.

REPRODUCTION: *D. millepunctatus* is assumed to be an egg-layer. Within the genus *Diploglossus*, most species are livebearers; the reproductive behavior of *D. millepunctatus* has not been verified. Gravid females and a copulating pair were observed in March during the peak of the booby's nesting season. It seems probable that the Malpelo Galliwasp schedules its reproductive cycle with booby nesting and the availability of high-value food resources.

MISCELLANEA: This lizard is the most abundant species of three endemic lizards. It is strongly diurnal with its visible abundance increasing through out the morning and then declining as individuals return to rock crevices and other hiding places by noon.

These lizards are opportunistic carnivores. They eat the usual invertebrate prey of lizards, and during the booby-nesting season, they eat fish dropped when adult boobies are feeding their nestlings as well as booby feces; they also feed off seaside carrion.

Varanidae: Monitors and Goannas

Varanids are an exclusively Old World group of predominantly tropical and subtropical lizards occurring throughout Africa, southern Asia from the Mediterranean coast to southern China, and south to and through most of Australia. Presently, more than 70 species are recognized, and all are contained within the genus *Varanus*. A dozen or so of these monitors have broad distribution, including *Varanus niloticus* throughout much of Africa and *V. salvator* from India and Sri Lanka through Southeast Asia to Sulawesi. All monitors are carnivores, and most are active foragers that search for their prey by sight and smell, having a snake-like long, forked tongue. They vary in size from the Pygmy Goanna at 120 mm SVL to the Komodo Dragon at ~1.8 m SVL.

Varanus "indicus"   **Pacific Monitor**

Plate 26

APPEARANCE: Large, greyhound-like lizard, with slender body and long narrow head atop long, sturdy neck; a robust body; short, strong limbs; and long, sturdy tail, laterally compressed on posterior half. Dorsal ground color is black from tip of snout to tip of tail, and small spots over entire dorsal surface; spots are various light colors, depending upon locality, and always composed of one to four light-colored scales. Venter from chin onto trunk has whitish to light-tan ground color, darkened by diffuse black marks and mottling. Neck, trunk, limbs, and tail have covering of small, smooth, circular to elliptical scales, largely arranged in transverse series; scales are in juxtaposition and equal in size dorsally and laterally, and larger (two times) ventrally. Scales on head are small, smooth, abutting plates of various sizes but none greatly enlarged. Middorsal scales are in about 140 to 150 rows from nape to base of tail, approximately 110 to 120 scales encircle body in middle of trunk, and fourth toe (hindfoot) bears about 25 scale rows on its underside.

SIZE: Females are smaller than males; further, regional differences in body size might be present although data have not been presented statistically to test such differences. In Guam, adult females range from 275 to 440 mm SVL; males from 320 to 580 mm SVL. Tail length is about 120% to 150% of SVL. Hatchling size is about 270 mm SVL.

OCCURRENCE: The Pacific Monitor occurs spottily in Micronesia. It is widespread, although usually not abundant, in the major islands of Palau. Populations are present in Yap, Ulithi, Ifalik, Pohnpei, Kosrae, Guam, and most of the Northern Mariana Islands. Elsewhere, it occurs in northern Australia and New Guinea, where it is largely coastal and is known as the Mangrove Monitor.

HABITAT CHOICE: Monitors occur in forested situations ranging from thick, brushy scrub arising from human disturbance to well-developed secondary forest.

REPRODUCTION: As with size at maturity and maximum size, reproduction of *V. indicus* has regional differences. For most populations, females produce small clutches of two to four eggs, although a clutch size of nearly 20 eggs has been reported. Eggs are deposited in nests dug in soil. Incubation varies from about 150 to nearly 200 days. In Guam, egg-laying appears seasonal, from September to December.

MISCELLANEA: Is this species introduced or native, and, if introduced, did it arrive with early Pacific colonizers or is it a post-European arrival? These questions have not been answered, although the limited zooarcheology of western Pacific islands reveals no pre- or immediately post-arrival evidence of *Varanus*. Crombie and Pregill noted that Palau and western Caroline populations might be native with these monitors arriving prior to human colonization. The eastern Caroline and Mariana populations are possibly post-European arrival into the western Pacific. There are coloration differences among the different island groups, although these differences have not been examined systematically.

OTHER LIZARDS, MOSTLY ALIENS

This assortment contains two pet-trade species that were intentionally released in the Hawaiian Islands and a supposed Palauan lizard, whose origin likely results from a mislabeled museum specimen.

Agamidae: Dragons and Relatives

The agamids are members of the Iguania clade. They are exclusively Old World in distribution, with four areas of species radiation: Africa, subtropical and temperate Southwest and Central Asia, South and Southeast Asia, and Australia. There are about 40 genera and over 400 species of agamid lizards. Few of the genera occur in more than one of the preceding geographic areas. Agamids have not dispersed into the Pacific with the possible exception of a single species in Palau.

Hypsilurus godeffroyi   **Papua Angle-headed Lizard**

Figure 18

Is the Palauan tree dragon an erroneous record, a surviving highly cryptic species, or a recently extinct species? Crombie and Pregill examined these alternatives, and—owing to a 1985 discovery of some subfossil remains in a gravel pit—with some hesitancy decided that this dragon may persist. Bauer and Watkins-Colwell examined the type of *H. godeffroyi* and documented the history of its type specimen and concluded that the type most likely derived from the north coast of New Guinea, not Palau.

APPEARANCE: Large, full-bodied lizard with slender, moderately long limbs and very long, compressed tail. Its head is large, especially in lateral view, with gular pouch nearly as large as head. Color data for living individual are unavailable; following coloration is from preserved specimens. Dorsal ground color light to medium brown; juveniles have four broad transverse bars from middorsal crest to midlaterally, another bar on neck and additional ones on anterior quarter of tail; bars are lost in adults, and few adults have broad middorsal area darker than sides. Venter is lighter brown. Aside from dorsal crest, neck and trunk

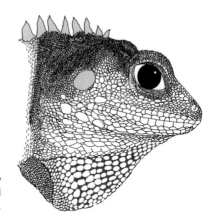

FIGURE 18. Head morphology of the Papua Angle-headed Lizard, *Hypsilurus godeffroyi*.

have small, trihedral scales, gradually becoming larger ventrally; ventral scales are large, strongly keeled, and imbricate. Dorsal crest is continuous from nape to middle of tail and higher in males than females, totaling 52 to 53 spines from nape to vent, nine to 10 cervical ones, then three to four small ones before large trunk ones. Head lacks plates dorsally, and instead is covered by small, imbricate, keeled scales; side of head with small smooth scales and only nasal, rostral, and lip scales are medium-sized and plate-like.

SIZE: Females attain a maximum SVL of 147 mm; and males range from 168 to 177 mm SVL. Tail length is 330% to 340% of SVL.

OCCURRENCE: This species, as presently defined, is a New Guinean lizard, principally found on the northern half of the island from the coast and nearshore islands to mid-montane level. The presumed Palauan record of a subfossil is from Ngeaur, but see comments in the Miscellanea section.

HABITAT CHOICE: Angle-heads are predominantly forest inhabitants.

REPRODUCTION: Aside from being an egg-layer, reproductive data are lacking.

MISCELLANEA: If a tree dragon persists or did exist in Palau, its scientific name will remain *H. godeffroyi* because the two syntypes derive from the "Pelew-Inseln." Of course, that name retention relies on present-day voucher specimens matching the morphology of the syntypes. There is strong evidence that these syntypes were mislabeled and did not derive from Palau, but rather are from islands off the northern coast of New Guinea.

Chamaeleonidae: Chameleons

Chameleons are like turtles. Once you have seen a chameleon, you will immediately recognize all other chameleons. The strongly compressed body that is two or more times higher than wide; the short, usually prehensile tail; the head with a bony casque over neck, muffler-like lids covering independent moveable eyes; and the zygodactylous, mitten-like feet are features that occur in combination in no other group of lizards. Chameleons are members of the Iguania clade and consist of 10 genera and nearly 200 species. They occur throughout Africa, except in the Sahara, edging into the Near East and southern Arabian Peninsula, and into peninsular India. All species have projectile tongues to snare their prey, which are mostly insects. In spite of an anatomy strongly adapted for an arboreal life style, they also live in a variety of habitats other than forest, such as arid scrub to grasslands. In these latter areas, they are terrestrial. The two species occurring in the Pacific are recent introductions.

Chamaeleo calyptratus   **Veiled Chameleon**

Plate 26

APPEARANCE: Large, laterally flatten lizard with robust limbs and short, prehensile tail. Head is large, laterally compressed with very large casque equal or exceeding the depth of head (sharkfin-like) on the rear of head and large gular pouch. In its bold coloration phase, ground color is turquoise green to yellowish green. Each side of trunk has four or five dark, black-and-green vertical bars. Head with black canthal stripes, black postorbital stripe from posterodorsal corner of eye to posterolateral base of casque, and casque has black vertical stripe on its anterior edge and another in middle, often more brightly colored, and lips dark and light barred. Narrow venter is green. Scalation is largely smooth granular scales, varying in size by location. Those on head are largest granular scales; those scales edging casque and other cranial ridges and lips are the largest scales on head, yet they are still small scales. Anterior crest scales of middorsal and midventral crests are small, conical, and erect, decreasing in size posteriorly. Scales of trunk are of uniform size from dorsal to ventral crest and tightly packed; midlaterally and ventrolaterally, scales form elliptical clusters outlined by very small granular scales and superficially producing appearance of larger scales.

SIZE: Females appear to average smaller than males but comparative measurement data are not available. Adult females are about 180 mm SVL; males about 210 mm SVL. Hatchling size has not been reported.

OCCURRENCE: The Veiled Chameleon was introduced in Maui in the 1990s by hobbyists, who continue its spread throughout the islands. It current distribution remains uncertain because of the preceding human behavior. Its native distribution is small and includes Yemen and southern edge of Saudi Arabia.

HABITAT CHOICE: Presently this chameleon is confined to horticultural landscapes; however, it has a broad ecological tolerance and can live and reproduce in habitats from dry, sea-level scrub forest to very wet montane forest. It is this adaptability to diverse habitats, its large size, and its catholic diet that makes this alien lizard a dangerous predator of endemic Hawaiian animals from invertebrates to birds.

REPRODUCTION: Veiled chameleons lay eggs, unlike the live-bearing Jackson's Chameleon. A female can lay 30 to 95 eggs in a nest that she digs in the soil. A female is reported to be capable of laying three clutches a year and reproducing year-round, although these features of its reproduction remain unconfirmed for the Hawaiian Islands. Incubation is about six months.

MISCELLANEA: Veiled Chameleons are primarily insectivores, although they also eat leaves, flowers and buds, small mammals, and birds.

Trioceros jacksonii   **Jackson's Chameleon**
Plate 26

APPEARANCE: Midsize to large, laterally flattened lizard with robust limbs and short, prehensile tail. Head is large, laterally compresses with small casque on rear of head; males have three horns on head, modest-length preorbital pair projecting horizontally on each side in front of eye, and larger, median horn on snout; adults females commonly lack horns or occasionally have single, small rostral one; gular pouch is absent. In its bold coloration phase, ground color is medium green. Each side of trunk has four or five dark, black-and-green vertical bars. Head is green with yellow-bordered casque and tan horns and lips. Lower sides of trunk and venter are uniform green to greenish white. Scalation is largely smooth granular scales of variable and mixed sizes, yet all are relatively small. Head scales are some of largest granular scales, especially those on the dorsal surface from snout to base of casque. Male's horns are modified scales. Dorsal crest is series of separate spine clusters, 16 to 18 clusters from posterior edge of neck to sacrum; each cluster has pair of laterally flattened, sharp-tipped conical scales, posterior one of each pair largest. There is no midventral crest. Trunk scales from dorsal crest to ventrolateral surface are mixture of sizes and shapes, all of which are smooth and slightly tuberculate; some are true tubercles and are widely scattered among smaller scales of neck, trunk, and limbs. Venter has more uniform scalation of slight tuberculate granular scales.

SIZE: Adult males are reported to be significantly larger than females, yet data from the Hawaiian populations suggest otherwise. Adult females range from 90 to 130 mm SVL (mean, 107 mm SVL); males from 70 to 140 mm SVL (98 mm). Because males mature at a smaller size (70 mm SVL) in the Hawaiian Island populations compared to native populations (minimum mature males 90 mm SVL in Kenya), it is possible that the Hawaiian population is reproducing at a younger age, and the slower growth of adults yields an overall size reduction in this alien population. The precise geographic origin of the Hawaiian population is not known. Different populations (subspecies) of *T. jacksonii* have very different average SVL, ranging from the smallest at 78 mm to largest at 124 mm.

OCCURRENCE: Jackson's chameleon now occurs on all main Hawaiian Islands except Kaua'i. It is an exotic species and became established on O'ahu in the early 1970s. Its original release and subsequent illegal introduction throughout the islands has been by reptile hobbyists and aficionados, who mistakenly believe that the introduction of an exotic animal enhances nature and has no toxic biological effects. The native distribution of this chameleon is central East Africa (Kenya and Tanzania).

HABITAT CHOICE: In the Hawaiian Islands, this chameleon has become widespread from urban-suburban horticultural landscapes to native dry forest and high-elevation native wet forest.

REPRODUCTION: Jackson's Chameleon is a livebearer and produces litters of seven to 21 neonates in O'ahu. A report of Hawaiian litters of up to 50 neonates has not been confirmed for wild populations. Data suggest that this chameleon has become an aseasonal breeder in the Hawaiian Islands, although it is distinctly a seasonal breeder in Africa; however, in Africa, it occurs at a much higher elevation and a more seasonal environment.

MISCELLANEA: Jackson's Chameleon, like all chameleons, is a stealth predator. Hidden by its coloration and body shape, it moves slowly to within "tongue's reach," and then projects its tongue to capture its prey. A recent study showed that *T. jacksonii* searches for prey through a steady, very slow walking gait; the authors labeled this behavior "cruise foraging."

In the Hawaiian Islands, *T. jacksonii* eats primarily insects, with dipterans (flies) as its major prey, closely followed by homopterans and orthropterans. The diet of individuals living in a native dry forest included nearly 40% endemic species. These data highlight the threat of this invasive chameleon to native Hawaiian insects and other invertebrates.

SNAKES

Snakes are uncommon residents of Pacific islands, and land-living snakes are often uncommon or inconspicuous. Terrestrial species are modestly diverse, representing five families of snakes: blindsnakes, boas, colubrids, elapids, and homalopsids. The evolution of unique species on distant islands attests to the arrival of some snakes well prior to arrival of humans in the Pacific. All but one of the Pacific terrestrial snakes are nonvenomous, and the venomous one—the Bola (Elapidae) of Fiji—is small and disinclined to bite; however, a word of caution: seakraits spend considerable time ashore and often are abundant in the above-tide areas in some coastal areas. The other venomous snakes are the seasnakes and seakraits, also elapids and generally disinclined to bite, but they possess strong and potentially deadly (for humans) venoms.

Typhlopidae: Blindsnakes

All blindsnakes are burrowers. Most spend their entire lives within the ground or at the soil-litter interface; surprisingly, individuals are occasionally found climbing a tree or in an epiphyte. Their burrowing adaptations include a surface of smooth, tightly overlapping scales on a cylindrical body of nearly equal diameter from the blunt or slightly pointed head to a short, abruptly tapering tail. Eyes are visible on most species and lie beneath large head scales; presumably the eyes function to perceive the presence or absence of light. Blindsnakes prey on soft-bodied invertebrates; they have a preference for termites and ants, and their eggs and larvae. The smooth, curved body surface protects them from the biting mandibles of the soldier termites and ants. Blindsnakes and their scolecophidian relatives are an ancient divergence from the main snake lineage and appear to be structurally quite primitive. In spite of this appearance, they are a specialized and successful snake group with more than 300 species that occur throughout the world's tropics.

Ramphotyphlops acuticaudus   **Palau Blindsnake**

Plate 27

APPEARANCE: Small, shiny brown snake. This pencil-diameter snake is brown above, grading ventrolaterally into yellowish-brown venter. Head is short with bluntly rounded snout; head indistinguishably joins cylindrical body that is equal diameter throughout its full length and imperceptibility becomes short, spine-tipped tail. Body scales are smooth, imbricate, and occur in 22 rows at midbody; there are 306 to 393 dorsal scale rows, and ventral scales are identical in size and shape to dorsal scales.

SIZE: Adults range from 152 to 243 mm SVL with tail length 3% to 5% of SVL.

OCCURRENCE: This species is endemic to Palau and lives on volcanic Babeldaob, Ngeaur, Ngercheu, and Oreor.

HABITAT CHOICE: The Palau Blindsnake occurs widely from the forest edge into town gardens.

REPRODUCTION: Reproductive data are not available; *R. acuticaudus* presumably lays eggs.

MISCELLANEA: The abundance of *R. acuticaudus* seems to be declining as the Brahminy Blindsnake population increases. The Brahminy Blindsnake probably arrived in Palau, specifically Oreor, during World War II or shortly thereafter. It is gradually spreading through the islands.

Another blindsnake was recently found on the limestone islands south of the distribution of *R. acuticaudus*; it probably represents a new species.

Ramphotyphlops adocetus   **Ant Atoll Blindsnake**
Plate 27

APPEARANCE: Small or barely midsize, shiny brown snake. This pencil-diameter snake is bicolor, with brown above grading quickly ventrolaterally into pigment-free venter. Head is short with wedge-shaped snout; head indistinguishably joins a cylindrical body that is equal diameter throughout its full length and imperceptibility becomes short, spine-tipped tail. Body scales are smooth, imbricate, and occur in 22 rows at midbody; there are 447 to 474 dorsal scale rows from rostral scale to tip of tail. Ventral scales are identical in size and shape to dorsal scales.

SIZE: The few known specimens of this blindsnake range from 154 to 390 mm total length. Tail length is 4% to 5% of total length.

OCCURRENCE: This species is known only from a single islet (Pasa Island) in the Ant Atoll of the Caroline Islands, Federated States of Micronesia.

HABITAT CHOICE: All individuals of this blindsnake have been found beneath decaying plant detritus or, in most instances, within rotten coconut logs. It often occurs in small aggregations.

REPRODUCTION: Reproductive data are not available *R. adocetus* presumably lays eggs.

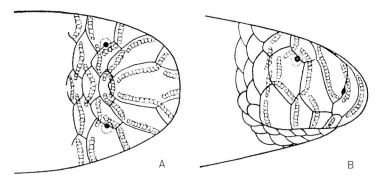

FIGURE 19. Head morphology of the Brahminy Blindsnake, *Ramphotyphlops braminus*. (A) Dorsal view. (B) Lateral view. The solid black lines denote scale margins; the lines of circles identify the location of series of glandular scale organs.

MISCELLANEA: The presence of this species on a single islet within the Ant Atoll group and the absence of rats on Pasa Island suggest that rats are its major predator and the cause of its absence from other islands in the atoll cluster.

Ramphotyphlops braminus   **Brahminy Blindsnake**

Plate 27, Figure 19

APPEARANCE: Small, shiny black snake, smallest of all Pacific species. This spaghetti-diameter snake is black to dark brown above, grading ventrolaterally into grayish-brown venter. Head is short with bluntly rounded snout; head indistinguishably joins cylindrical body that is equal in diameter throughout its full length and imperceptibility becomes short, spine-tipped tail. Body scales are smooth, imbricate, and occur in 20 rows at midbody; there are 261 to 358 dorsal scale rows from rostrum to tip of tail. Ventral scales are identical in size and shape to dorsal scales.

SIZE: Adults range from 98 to 180 mm total length (95–175 mm SVL). The tail is 2% to 3% of total length. Hatchlings are 53 to 62 mm total length.

OCCURRENCE: The Brahminy Blindsnake occurs widely among the Pacific islands, but it is often undetected owing to its small size and worm-like appearance. Its origins are uncertain, possibly India. It has become established worldwide in the tropics because of its transport in the root-balls of nursery plants, hence its vernacular name "Flowerpot Snake."

HABITAT CHOICE: Once established on an island, it becomes ubiquitous.

REPRODUCTION: *R. braminus* is an all-female species that reproduces parthenogenetically. It lays small clutches of one to eight tiny oblong eggs. Incubation takes about eight weeks. In some areas, reproduction is seasonal—for example, reproduction occurs from May to July in the Ryukyu Islands. Elsewhere egg production may occur year-round, although no data are yet available to support this hypothesis.

MISCELLANEA: Its diet consists of soil invertebrates; possibly it has a predilection for the eggs and larvae of termites and ants.

Ramphotyphlops depressus　　　　　　　　**Melanesia Blindsnake**

Plate 27

APPEARANCE: Small, shiny brown snake. This pencil-diameter snake is light to dark yellowish brown or orangish brown above, grading ventrolaterally into pale-yellow to light-brown venter. Head is short with bluntly rounded snout; head indistinguishably joins cylindrical body that is equal in diameter throughout its full length and imperceptibility becomes short, spine-tipped tail. Body scales are smooth, imbricate, and occur in 22 to 24 rows at midbody; there are 289 to 438 dorsal scale rows from rostrum to tip of tail. Ventral scales are identical in size and shape to dorsal scales.

SIZE: Adult total length ranges from 180 to 220 mm (mean 208 mm). The tail is about 2% to 3% of total length.

OCCURRENCE: *R. depressus* occurs from the Bismarck Archipelago, Papua New Guinea into and throughout the Solomon Islands.

HABITAT CHOICE: This is a forest snake, found beneath the forest-floor litter.

REPRODUCTION: Reproductive data are not available. *R. depressus* presumably lays eggs.

Ramphotyphlops hatmaliyeb　　　　　　　　**Ulithi Blindsnake**

Plate 27.

APPEARANCE: Small to midsize, shiny brown snake from above. This pencil diameter snake is bicolor, brown above and grading quickly ventrolaterally into pigment-free venter. Head is short with wedge-shaped snout; head indistinguishably joins cylindrical body that is equal in diameter throughout its full length and imperceptibility becomes short, spine-tipped tail. Body scales are smooth, imbricate, and occur in 22 rows at midbody; there are 452 to 472 dorsal scale rows from rostral scale to tip of tail. Ventral scales are identical in size and shape to dorsal scales.

SIZE: The few known specimens of this blindsnake range from 178 to 416 mm total length. The tail length is 3% to 4% of total length.

OCCURRENCE: This species lives on several of the islets comprising Ulithi in the Yap Islands, Federated States of Micronesia.

HABITAT CHOICE: *R. hatmaliyeb* is a burrower and is typically found beneath plant debris in forest and scrub habitats. However, like some other species of blindsnakes, it forages on the surface at night and uses bark crevices to climb trees.

REPRODUCTION: Reproductive data are not available; *R. hatmaliyeb* presumably lays eggs.

Ramphotyphlops nsp   Taveuni Blindsnake
Not illustrated

APPEARANCE: Small, shiny brown snake. This pencil-diameter snake is brown above, lighter brown ventrolaterally and grading into brownish-yellow venter; lighter brown gives impression of light stripe on each side. Head is short with bluntly rounded snout; head indistinguishably joins cylindrical body that is nearly equal diameter throughout its full length and imperceptibility becomes short, spine-tipped tail. Eye with pupil is visible below head scale. Body scales are smooth, imbricate, and occur in 22 (rarely 21) rows at midbody; there are 305 to 340 dorsal scale rows; scales are nearly equal in size around body.

SIZE: The known body sizes range from 104 to 249 mm SVL; adults are probably 200 mm SVL or larger. The tail is short, 2% to 3% of SVL.

OCCURRENCE: This blindsnake is known only from Taveuni, Fiji.

HABITAT CHOICE: Much of Taveuni is forested. This blindsnake appears to be confined to the forest margin and grassland on the southern end of the island.

REPRODUCTION: Reproductive data are not available; *Ramphotyphlops* nsp presumably lays eggs.

MISCELLANEA: Its diet is likely to consist of soil invertebrates, along with their eggs and soft-bodied larvae.

The presence of a Fijian (Taveuni) blindsnake was first reported in 1897 at a meeting of the Linnean Society of New South Wales. Two specimens were placed in the Australian Museum (Sydney) but inexplicably disappeared shortly thereafter. Although naturalists periodi-

cally searched for the blindsnakes, it was a century before local Taveu-
nians reported a location where the blindsnakes were common. New
specimens of this snake were obtained, confirming its existence and its
status as an undescribed species.

Boidae: Boas

Boas are distributed worldwide, although they are absent from much of
sub-Saharan Africa and southern Asia where one or more python species occur. Boas are only moderately diverse, with about three dozen
species. A few of the species, such as Anacondas (total length to 11 m),
are quite large, although most are more modest in size—less than 3 m
in total length; a few species attain lengths less than 1 m. All boas are
constrictors, and many prey exclusively as adults on birds and mammals. All boas are livebearers, in contrast to the egg-laying pythons.

Candoia bibroni   **Pacific Treeboa**

Plate 28, Figure 20

APPEARANCE: Midsize to large snake with variable dorsal pattern of middorsal zigzag stripe of linked, light-edged brown blotches. Dorsal ground
color varies from light tan and orange to medium gray, or uncommonly
nearly black; linked blotches are several shades darker and light border
is double streak of cream and rufous brown. Laterally and ventrolaterally, trunk bears numerous smudged brown blotches that extend onto
ventral scales, sometimes narrowly touch midventrally; venter also has
numerous dark-brown spots and mottling on cream to tan background.
Tail is usually dark below. Head is elongated triangle in dorsal outline
with square snout and distinctly delineated from neck. In lateral view,
top of snout projects beyond mouth, resulting in slightly beveled snout.
Body is cylindrical and narrow anteriorly, becoming robust and strongly
laterally compressed by midbody through to tail. Tail is of moderate
length and ends with blunt spine. Head scales are medium-sized and
regularly arranged. Eye separated from supralabial scales by row of
small suborbital scales. Dorsal trunk scales are lightly but distinctly
keeled except for the ventral-most one to four rows. Dorsal scale rows
are 25 to 35 anteriorly, 29 to 43 at midbody, and 19 to 24 posteriorly.
Ventral scales number 203 to 206. Males and most females have pelvic
spurs, horny spine on each side of vent, larger in males.

SIZE: Females average longer than males. Adult females range from 600
to 1,460 mm SVL (mean 983 mm SVL); males from 460 to 1,190 mm

SVL (720 mm). The tail is 12% to 17% of SVL. Neonates range from 160 to 230 mm SVL.

OCCURRENCE: *C. bibroni* is a Southwest Pacific species, occurring from the eastern Solomon Islands (San Cristobal) through Vanuatu and Loyalty Island (possibly an introduction) to Fiji and Samoa.

HABITAT CHOICE: The Pacific Treeboa is strongly arboreal and occurs in a variety of forested habitats from cacao plantations and degraded secondary forest to primary forest.

REPRODUCTION: It is a livebearer and gives birth to litters of two to 33 neonates (mean ~6). The size of the litter has a strong association with female size; larger females average more young per litter.

MISCELLANEA: *C. bibroni* preys on both ectothermic and endothermic vertebrates. Younger individuals depend largely on lizards; adults prey mainly on birds and mammals.

Subtle skeletal differences hint that *C. bibroni* might consist of two species, a central Pacific one and a Vanuatu-Solomon Islands one.

Candoia paulsoni   **Melanesia Bevel-nosed Boa**

Plate 28

APPEARANCE: Large snake with middorsal zigzag stripe of linked broad blotches. Dorsal ground color varies from light tan to medium brown; linked blotches are several shades darker brown, occasionally with lateral edges tipped in dark brown and typically with small, light brown spots within the blotches. Laterally and ventrolaterally, trunk bears numerous brown blotches that end on the border of the ventral scales; venter has numerous dark-brown spots and mottling on cream to tan background. Tail is white to cream beneath large dark-brown blotches; the first lies immediately behind the vent. Head is elongated triangle in dorsal outline with square snout and distinctly delineated from neck. In lateral view, top of snout projects beyond mouth, resulting in beveled snout. Body is cylindrical and narrow anteriorly, becoming robust and strongly laterally compressed by midbody through to tail. Tail is strongly prehensile, moderately long, and ends with a blunt spine. Head scales are medium-sized and regularly arranged. Eye not separated from supralabial scales by row of subocular scales. Dorsal trunk scales are lightly but distinctly keeled except for ventral-most two or three rows. Dorsal scale rows are 27 to 35 anteriorly, 34 to 42 at midbody, and 22 to 28 posteriorly. The ventral scales number 184 to 202. Males and most females have pelvic spurs, horny spine on each side of vent.

SIZE: Females average longer than males. The maximum total length of females is 1,365 mm; males 840 mm. Tail is 11% to 18% of SVL length. Neonates range from 190 to 240 mm total length.

OCCURRENCE: This boa just edges into the Pacific herpetofauna of this guide. It is reported to occur on Santa Cruz the largest island of the Santa Cruz group. Elsewhere, it is found westward throughout the Solomon Islands to Bougainville, disjunctly in northern and eastern Papua New Guinea and Halmahera area.

HABITAT CHOICE: This boa lives in both secondary and primary forest.

REPRODUCTION: Information on this species reproduction is limited. It is a livebearer, producing litters of 16 to 27 young.

MISCELLANEA: Generally, *C. paulsoni* is an excitable snake when disturbed. It flees or forms a ball of coils and will strike quickly from the balled posture. Its prey is mainly birds and mammals.

C. *paulsoni* is believed to have six geographic races or populations; the preceding data apply only to the eastern-most population

---

Candoia superciliosa   **Palau Bevel-nosed Boa**

Plate 28, Figure 20

APPEARANCE: Midsize snake with middorsal stripe of linked broad blotches. Dorsal ground color varies from tan to medium brown; linked blotches are several shades darker brown, often with lateral edges tipped in dark brown. Laterally and ventrolaterally, trunk bears numerous grayish-brown blotches that overlap onto ventral scales, giving venter spotted pattern on a cream to tan background laterally. Tail is brown and white beneath with elongate white, dark-bordered spot immediately behind vent. Head is elongate triangle in dorsal outline with square snout and distinctly delineated from neck. In lateral view, top of snout projects beyond mouth, resulting in a beveled snout. Body is cylindrical, anteriorly narrow becoming robust by midbody through to tail. Tail is strongly prehensile, moderate length, and ends with blunt spine. Head scales are medium-sized and regularly arranged. Eye touches supralabial scales. Dorsal trunk scales are lightly but distinctly keeled except for ventral-most two or three rows. Dorsal scale rows are 24 to 28 anteriorly, 31 to 36 at midbody, and 20 to 25 posteriorly. Ventral scales number 165 to 192. Males and most females have pelvic spurs, horny spine on each side of vent.

SIZE: Females average longer than males. The maximum total length is 885 mm for females and 546 mm for males. Most Palau adults range

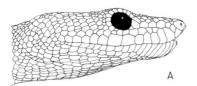
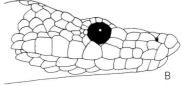

A

B

FIGURE 20. Head morphology of Pacific Bevel-snout Boas. Lateral views of *Candoia bibroni* (A) and *Candoia superciliosa* (B). The former species has subocular scales, and the eye is separated from the lip scales; the latter species and *C. paulsoni* lack subocular scales and the eye touches the lip scales. The large dorsal humps on the rear of the heads are the protrusion of the massive jaw muscles.

from 400 to 600 mm SVL. Tail is 14% to 20% of SVL, averaging ~18%. Neonates range from 160 to 230 mm SVL.

OCCURRENCE: *C. superciliosa* is a Palauan endemic and is found widely throughout the main islands.

HABITAT CHOICE: This Bevel-nosed Boa has adapted well to human landscape modification and occurs from house-side shrubbery and gardens into the forest. Although it is an arboreal snake, feeding and sleeping in trees and shrubs, it also regularly forages on the ground.

REPRODUCTION: Knowledge of this boa's reproductive behavior is limited. It is a livebearer and produces litters of 16 to 27 neonates.

MISCELLANEA: *C. superciliosa* is an exceptionally docile snake and does not attempt to bite when handled.

Colubridae: Common Snakes

The Colubridae is the largest group of snakes comprising most non-fanged (nonvenomous) and rear-fanged snakes of the Americas, Euroasia, northern Africa, Asia, and Australia. Many non- and rear-fanged snakes of sub-Saharan Africa were considered part of the colubrid lineage until recently, but these species are now classified in a separate family. The colubrids are incredibly diverse in size, morphology, behavior, and ecology. Species range in size from tiny species such as the North American Worm Snakes (at 19–28 cm SVL) to giants such as the Indian Ratsnake (which can reach 3.5 m SVL). Diet is similarly variable and spans the available organisms, ranging from earthworms and soil invertebrates to warm-blooded birds and mammals. Some species are burrowers, others arboreal or aquatic. Appearance also varies from short, stout species to long, slender ones. Some colubrids lay eggs and others are livebearers. The colubrid evolutionary radiation has been very successful, and more than half of the nearly 3,000 living species of snakes are colubrids.

Alsophis biseralis   **Galápagos Racer**
Plate 29

APPEARANCE: Midsize to large snake with variable patterns of cream to medium-brown stripes and spots to nearly uniform on dark-brown ground color. Ground color of venter is light yellow to white; however, venter usually appears dusky owing to heavy brown mottling or spotting. Head is small, obovate in dorsal outline, with blunt snout and slightly wider posteriorly than neck. Body is slender and cylindrical, and has long slender tail. Body scales are smooth, overlapping, and occur in 19 rows at midbody with 199 to 250 ventral scales.

SIZE: Adult males and females appear equal in size. Adult females attain lengths of 872 to 945 mm SVL in the different insular populations (483 to 945 mm SVL range for all populations); males range from about 513 to 1,005 mm SVL. Tail lengths of adult females are 26% to 35% of SVL; tail lengths of males are 28% to 35%.

OCCURRENCE: This Racer occurs more broadly in the Galápagos than any of the other terrestrial snakes. Populations occur on the islands of Baltra, Bartolome, Fernandina, Floreana, Isabela, Rábida, San Cristóbal, Santa Cruz, Santa Fé, and Santiago.

HABITAT CHOICE: The Galápagos Racer is a terrestrial species of the various arid to mesic habitats on its various home islands.

REPRODUCTION: Reproductive data for this egg-laying species have not been documented.

MISCELLANEA: This species currently consists of three populations that differ in color pattern and adult size.

Antillophis slevini   **Galápagos Banded Snake**
Plate 29

APPEARANCE: Midsize snake with dark-brown bands on yellowish background. Dark bands do not encircle body, but end above ventral scales. Ground color of venter is light yellow to white; rear edge of each ventral scale is dark-edged. Head is small, obovate in dorsal outline, with blunt snout and slightly wider posteriorly than neck. Body is slender, cylindrical, and has long slender tail. Body scales are smooth, overlapping, and occur in 19 rows at midbody; there are 170 to 184 ventral scales.

SIZE: Measurement data are limited for *A. slevini*. An adult female measured 413 mm SVL. A small sample of adult males range from 345 to 375 mm SVL (mean 360 mm). Tail length is 39% to 47% of SVL in adult males.

OCCURRENCE: The Galápagos Banded Snake occurs on three islands: Fernandina, Isabela, and Pinzón.

HABITAT CHOICE: This snake is terrestrial.

REPRODUCTION: No reproductive data are available for this egg-laying species.

Antillophis steindachneri   **Galápagos Striped Snake**

Plate 29

APPEARANCE: Midsize snake with dark-brown background and broad, cream dorsolateral stripes. Venter is alternating blocks of white and dark brown; each ventral scale is dark on its anterior and posterior edges. Head is small, short obovate in dorsal outline, with blunt snout and barely wider posteriorly than neck. Body is slender and cylindrical with a long tail. Body scales are smooth, overlapping, and occur in 19 rows at midbody; there are 166 to 184 ventral scales.

SIZE: Adult males average slightly larger than females. Adult females range from 337 to 370 mm SVL (mean 358 mm SVL); males from 343 to 403 mm SVL (370 mm). Tail length is 34% to 44% of SVL in adult females and 40% to 52% in males.

OCCURRENCE: The Galápagos Striped Snake lives on the islands of Baltra, Rábidan, Santiago, and Santa Cruz.

HABITAT CHOICE: *A. steindachneri* is a terrestrial species.

REPRODUCTION: No reproductive data are available for this egg-laying species.

Boiga irregularis   **Brown Treesnake**

Plate 30

APPEARANCE: Large to very large arboreal snake with numerous brown angular transverse bars on light-brown background. These dark bars become progressively diffuse posteriorly. Head is uniform brown dorsally and bears dark-brown postorbital stripe from eye to anterior edge of neck. Venter is brownish white, usually without markings; occasionally posterior trunk and tail have tiny dark spotting, and subcaudals are edged in dark brown. Head is large, ovate in dorsal outline, with blunt snout, large protruding eyes, and sharply delineated from neck. Body is slender, oblong, and flat-bottomed in cross-section, with long slender tail. Body scales are smooth, overlapping, and occur in 17 to 25 rows at midbody; there are 217 to 286 ventral scales.

SIZE: Adult males average substantially larger than females. Most adult females range from 84 to 165 cm SVL (mean ~108 cm); males from 80 to 240 cm SVL (~ 131 cm). Tail length is 20% to 22 % of SVL. Maximum total length for a Guam female is 199 cm and 310 cm for a male. Hatchlings are about 330 mm SVL.

OCCURRENCE: *B. irregularis* occurs throughout Guam and Saipan. Its native range extends from Halmahera through New Guinea into the Solomon Islands and northern Australia from the Kimberly to northern New South Wales.

HABITAT CHOICE: The Brown Treesnake occurs in all habitats in Guam, from urban landscaping through agricultural lands and savannas to the various forest types. A density of 16 treesnakes per hectare was estimated for a forest site in the mid-1980s when the populations were at their most dense. The Brown Treesnake remains abundant, although its average density appears lower. Although arboreal, these snakes regularly seek prey on the ground during periods of low moonlight intensity and revert to canopy search under bright moonlight conditions.

REPRODUCTION: Guam *B. irregularis* reproduces year-round in Guam. Males and females attain sexual maturity between 80 and 110 cm SVL. Females produce clutches of two to nine eggs (average clutch is four). Eggs appear to be laid in deep subterranean crevices; incubation is likely to require about six weeks.

MISCELLANEA: We know more about the biology and ecology of these Guam treesnakes than we know about any other terrestrial reptile in the Pacific. These arboreal snakes gained their well-deserved notoriety by their decimation of the Guam bird populations in the 1980s, thereby alerting us to the dangers of this alien species spreading elsewhere in the Pacific. In addition, their propensity for climbing electric utility structures leads to regular and costly electrical outages throughout the island. The need for population control or even, if possible, elimination has stimulated and financed numerous biological studies that continue today. The disruption of Guam ecosystems by these snakes and the economic costs they engender emphasize the need to prevent their spread to other Pacific islands.

Dendrelaphis salomonis   **Solomon Treesnake**

Plate 30

APPEARANCE: Large snake with variable coloration; dorsal ground color from grayish green or olive to light or dark brown; venter pale yellow on chin and anterior throat, posteriorly yellow to gray. Top of head is unicolor; diffuse irregular dark diagonal bars, separated middorsally, occur on neck and anterior-most trunk. Laterally, black stripe extends from snout through eye onto anterior trunk, sharply separating darker top of head from lighter lips and underside. Head is small, ovate in dorsal outline, with rounded snout and distinct from neck. Body is slender, loaf-shaped with long slender tail. Body scales are smooth, imbricate, and occur in 13 rows at midbody, with 171 to 211 ventral scales. Middorsal scales are enlarged, about 1.5 times larger than adjacent scales. Ventral scales are sharply folded ventrolaterally, creating flat venter with outer edges of each ventral scale turned upward and forming part of trunk wall.

SIZE: Measurement data are limited for this species. Adults average about 1 m in total length. Adults are probably sexually dimorphic, with females averaging about 10 mm longer than males. The tail is 40% to 50% of SVL. Data from a related species in tropical Queensland, *D. punctulatus*, confirms sexual dimorphism with female mean SVL of 1,012 mm (maximum 1,473 mm) contrasting with male mean of 929 mm SVL (1,202 mm).

OCCURRENCE: *D. salomonis* is widespread in the Solomon Islands, and is distributed from Bougainville (Papua New Guinea) to the Santa Cruz Islands.

HABITAT CHOICE: This species occurs widely from forest to garden. It searches for prey on the ground and in shrubs and trees. It is sight hunter and preys mainly on skinks.

REPRODUCTION: Detailed reproductive data are not available for this egg-laying species. *D. salomonis* probably is a seasonal breeder producing small clutches of four to 12 eggs, similar to *D. punctulatus*.

Dendrelaphis striolatus   **Palau Treesnake**

Plate 30, Figure 22

APPEARANCE: Midsize to large snake with variable coloration; dorsal ground color from dark blue to grayish green or olive, and venter from pale blue to yellow. Top of head is unicolor or with dusky smudge markings posteriorly; dark diagonal bars occur laterally on trunk, strongly defined anteriorly and becoming progressively diffuse posteriorly and disappearing in many specimens. Upper and lower lips are white extending from snout below eye to rear of head. Head is small, ovate in dorsal outline, with rounded snout and distinct from neck. Body is slender, loaf-shaped with long slender tail. Body scales are smooth, imbricate, and occur in 13, occasionally 11, rows at midbody, with 176 to 186 ventral scales. Middorsal scales are enlarged, about 1.5 times larger than adjacent scales. Ventral scales are sharply folded ventrolaterally, creating a flat venter with outer edges of each ventral scale turned upward and forming part of trunk wall.

SIZE: Adults range from 608 to 982 mm SVL (mean 724 mm). The tail is 40% to 50% of SVL.

OCCURRENCE: The Palau Treesnake is widespread in Palau, inhabiting all larger islands and many medium-sized ones.

HABITAT CHOICE: The species occurs across the entire spectrum of Palauan habitats from town shrubbery and gardens into forests. They are diurnal snakes, apparently terrestrial in prey searching but typically sleep in brushy fencerows, low shrubs, and in palm axils and fronds as much as 5 meters above the ground.

REPRODUCTION: They are egg-layers. Other details on reproduction have not been reported.

MISCELLANEA: The taxonomic status of this snake is uncertain. Is it a distinct species, a subspecies, or outlying population of *D. punctulatus*? I accept Crombie's and Pregill's interpretation of the Palauan population as a distinct species because of its widespread occurrence in Palau and the geographic divergence of color patterns within the Palauan archipelago.

Masticophis anthonyi   Clarion Racer

Figure 21

APPEARANCE: Large, slender-bodied snake with long tail. Dorsum is gray to brown with diagonal series of small black spots at each dorsal scale row (base of each dorsal scale is dark). Top of head is slightly darker than trunk and faintly marked with spots; eye is bordered anteriorly and posteriorly by white to cream bars; upper and lower lips also white to cream. Chin, throat, and anterior quarter of trunk usually

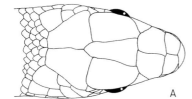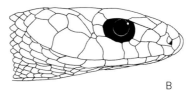

FIGURE 21. Head morphology of the Clarion Racer, *Masticophis anthonyi*. (A) Dorsal view. (B) Lateral view.

densely covered with dark spots; venter is beige to cream, occasionally becoming dusky gray posteriorly. Head is modest-sized, elongate ovate in dorsal outline, with rounded snout, merging smoothly into the neck. Body scales are smooth, imbricate, and occur in 17 rows at midbody; there are 186 to 204 ventral scales.

SIZE: Adults range from about 800 to 1,104 mm SVL, ~1,080 to 1,460 mm total length. The tail is 30% to 35% of SVL.

OCCURRENCE: This Racer occurs only on Clarion Island.

HABITAT CHOICE: *M. anthonyi* occurs throughout the island from the beaches to top of the island's central high point. When seabirds are nesting, Clarion Racers are frequently seen in the nesting colony, and are reported to eat nestlings, Clarion Treelizards, and hatchling sea-turtles. They forage in the open and grassy areas, but, when pursued, flee to rocky and cactus patches; the latter sites are also their preferred resting areas. Although strongly terrestrial in their activities, they occasionally bask on top of low cacti.

REPRODUCTION: No reproductive data have been reported for this egg-laying species.

Philodryas hoodensis   **Española Racer**

Plate 29

APPEARANCE: Midsize snake with pair of white dorsolateral stripes from eye to posterior third of trunk on dark background. Venter is also white; this white extends from lower lip onto ventrolateral surface of neck, body, and tail; ventral scales often have dark spot on each side, especially on anterior half of body. Head is small, obovate in dorsal outline, with blunt snout, and is slightly wider posteriorly than neck. Body is slender and cylindrical with moderately long and slender tail.

Body scales are smooth, overlapping, and occur in 17 or 19 rows at midbody; there are 199 to 214 ventral scales.

SIZE: Adult females average greatly larger than males. Adult females range from 529 to 856 mm SVL (mean 723 mm); males from 468 to 556 mm SVL (529 mm). Tail length is 26% to 35% of SVL in adult females and 36% to 42% in males.

OCCURRENCE: *P. hoodensis* has the smallest distribution of the Galápagos terrestrial snakes. It lives only on Isla Española and nearby Gardner Island.

HABITAT CHOICE: The Española Racer occurs through the island.

REPRODUCTION: No reproductive data have been reported for this egg-laying species.

MISCELLANEA: In contrast to the other Galápagos snakes, *P. hoodensis* is a constrictor, grabbing its lizard prey with its mouth and rapidly throwing coils about the prey. Constriction kills, not by suffocation, rather by compression of heart thereby disrupting blood flow, especially to the brain, and causing the prey's rapid death.

Elapidae: Seasnakes and Relatives

The elapids are venomous snakes with fixed fangs in the front of the mouth. Their venoms (each species has a unique toxic mix) tend to be neurotoxic, in contrast to the hemotoxic venoms of vipers. Elapids comprise a diverse group of snakes and include American coralsnakes, cobras, kraits, and the full variety of Australian snakes ranging from small burrowing snakes to the large terrestrial Taipan and Brown Snakes. The seasnakes and seakraits are members of the same evolutionary clade as the Australian elapids; this close phylogenetic relationship has been recognized only recently. Formerly, seasnakes were placed in the subfamily Hydrophiinae and seakraits in the Laticaudinae; now all these marine snakes as well as the Australian terrestrial snakes are contained in the Hydrophiinae, which includes the small Fijian Bola. Seasnakes and seakraits come in a variety of sizes and shapes, from midsize (~ 50 cm total length) to very large (> 2 m total length). All share a nearly cylindrical body shape anteriorly that becomes increasingly compressed from side to side beginning at midbody, with the tail becoming a vertically flattened broad fin. In all species, the ventral scales or plates are narrowed, although still moderately wide in the seakraits (*Laticauda*), but nearly or totally indistinguishable from the lateral trunk scales in many seasnakes.

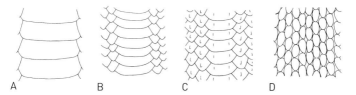

A   B   C   D

FIGURE 22. Shape and size of ventral scales in terrestrial and marine snakes. (A) Broad ventrals of the treesnake *Dendrelaphis striolatus*. (B) Reduced ventrals of semiterrestrial seakrait *Laticauda schisto-rhyncha*. (C) Further reduced ventrals in near-shore aquatic seasnake *Hydrophis coggeri*. (D) The ultimate in ventral reduction in the pelagic seasnake *Pelamis platura*.

Hydrophis coggeri   **Pacific Yellow-banded Seasnake**

Plate 31, Figure 22

APPEARANCE: Large seasnake with numerous (30–40) dark-olive to black bands on yellowish- to tannish-olive background, creating appearance of alternating, near equal-sized bands of dark and light encircling body. Head and neck are usually dark, occasionally with light areas on snout and nape. End of tail is totally black, and venter is dusky white. Head and anterior quarter of body are small and cylindrical, in contrast to thick and laterally compressed trunk and large laterally flattened tail. Body scales are knobbed or half keeled, barely imbricate, and occur in 29 to 34 rows at midbody; there are 280 to 360 narrow ventral scales. **SIZE:** Females are distinctly longer and heavier than males. Adult females range from 1,000 to 1,265 mm SVL; males rarely attain 1,000 mm SVL. The maximum reported total length is 1,364 mm (1,265 mm SVL). **OCCURRENCE:** In the Pacific, *H. coggeri* occurs around New Caledonia, Vanuatu, and Fiji. Elsewhere, it is found in the southern Indonesian waters to the Philippine Islands and Timor, and along the northern coast of Australia. **HABITAT CHOICE:** In the Pacific, this is a nearshore species commonly found in lagoons and near mangroves in areas with a soft or friable bottom and free-floating plant matter. It generally lives in shallow water several meters deep, although is found in depths up to 40 m. **REPRODUCTION:** All *Hydrophis* species are livebearers. *H. coggeri* appears to reproduce seasonally, from March to June in Fiji. Litter size varies from one to eight young.

MISCELLANEA: Most activity occurs at night. *H. coggeri* specializes in two families of eels (Ophichthidae, Congridae), which it captures by sticking its head down their burrows.

H. coggeri was regularly labeled *H. fasciatus* before it was recognized as a distinct species.

Hydrophis pacificus   **Pacific Faint-banded Seasnake**

Not illustrated

APPEARANCE: Large seasnake with 48 to 72 dark bands on a dark-gray background. Bands are black dorsally fading to obscure gray bands ventrally. Head has gray cap from midsnout to nape; its face from snout through eye to neck is lighter, with obscure pale flecks. Tail is banded as body, and underside is whitish. Head and anterior quarter of body are small and cylindrical, in contrast to thick and laterally compressed trunk and large, laterally flattened tail. Body scales are smooth, imbricate, and occur in 39 to 49 rows at midbody; there are 320 to 430 narrow ventral scales, barely larger than adjacent lateral scales.

SIZE: Adult *H. pacificus* has an average total length of 1,400 mm.

OCCURRENCE: *H. pacificus* has been reported as *H. belcheri* from the Solomon Islands and Caroline Islands. The latter locality seems unlikely; see comments below in the Miscellanea section. Its primary distribution is the Gulf of Carpentaria and the adjacent waters between Australia and New Guinea.

HABITAT CHOICE: In the Pacific, this species is a nearshore species commonly found in lagoons and near mangroves in areas with a soft or friable bottom and free-floating plant matter. Generally, these seasnakes forage in shallow water with a depth of several meters, although they have been found in depths up to 40 m.

REPRODUCTION: All *Hydrophis* species are livebearers. Data on the reproduction of *H. pacificus* are not available.

MISCELLANEA: Do any of the members of the genus *Hydrophis* occur widely in Oceania? Aside from *H. coggeri* in Vanuatu and Fiji, there is no evidence of any other resident populations in Oceania. Perhaps the reports of *Hydrophis* species in Micronesia (Carolines, *H. belcheri*) and Kiribati (Gilbert Islands, *H. ornatus*) represent waifs, although these records are more likely to represent misidentifications or mislabeled voucher specimens.

The identification of *Hydrophis* species has been difficult and is confounded by the lack of systematic studies. For example, *H. belcheri* now applies to populations of seasnakes in seas around Thailand. Before *H. pacificus* was recognized as a separate species, the populations between Australia and New Guinea were called *H. belcheri* or *H. melanocephalus*.

Laticauda colubrina   **Yellow-lipped Seakrait**

Plate 31

APPEARANCE: Midsize to large seasnake with 22 to 62 black bands on medium-blue to grayish-blue background, creating alternating bands of dark and light. All dark bands encircle body. Head has dark cap extending from eye posteriorly to first body band on neck; anterior half of head including the upper lip is cream to light yellow. Tail is banded as is body, and usually narrowly tipped in yellow. Underside is light yellow to cream from chin to tail. Head is robust and tapers slightly into robust cylindrical body that becomes laterally compressed on posterior quarter of trunk and strongly compressed tail. Body scales are smooth, overlapping slightly, and occur in 19 to 27 rows at midbody; there are 207 to 249 narrow ventral scales.

SIZE: Adult females average larger than males. Females range from 295 to 1,655 mm SVL (mean ~1,050 mm); males from 305 to 1,270 mm SVL (~700 mm). Tail length is modest, ~15% of SVL in females and ~18% in males.

OCCURRENCE: *L. colubrina* is a coastal species reported from Myanmar and Southeast Asia through the Sundas to northern Australia and eastward into the Pacific. Its Pacific distribution is spotty: it is found on Palau, eastern Micronesia, Samoa, Tonga, Fiji, Vanuatu, and the Santa Cruz Islands.

HABITAT CHOICE: The Yellow-lipped Seakrait is a shallow-water (usually < 25 m) inhabitant. It lives in both clear waters of reef and sand flats and in more turbid waters of river mouths and mangroves.

REPRODUCTION: *L. colubrina*, like all *Laticauda* species, is an egg-layer that must come ashore to deposit its eggs. In Fiji, it is a seasonal breeder, laying eggs from November into April. Clutch size ranges from three to 10 eggs (mean, six). Incubation is about 140 days.

MISCELLANEA: All species of *Laticauda* come ashore regularly, perhaps daily, and often spend many hours ashore. Seakraits are most active nocturnally and search for their fish prey, principally eels, then. The onshore behavior was assumed to be an adaptation to avoid snake-eating predators such as sharks and groupers. Recent physiological tests of *L. colubrina*, *L. laticaudata*, and *L. semifasciata* suggest an additional reason: rehydration. Seakraits do not drink saltwater, although they have salt glands; they apparently are unable to extract sufficient fresh water from their prey. When the snakes are ashore, they drink freshwater from rain pools or lap dew off vegetation. This behavior rehydrates the snakes and is probably a major driver for their regular terrestrial activity. During heavy rain storms, seakraits can drink from the freshwater lens on top of the seawater.

Laticauda frontalis   **Vanuatu Yellow-lipped Seakrait**

Plate 31

APPEARANCE: Midsize seasnake with black 30 to 40 bands on a medium-blue to grayish background. Anterior dark bands do not encircle body; some posterior bands fully encircle the trunk. Head has reduced dark cap on rear third of head; it is separated from first body band. Most of head, including upper lip, is cream to light yellow; dark postorbital stripe is small. Tail is banded as is body and is usually narrowly tipped in yellow. Underside is light yellow to cream from chin to tail. Head is robust and tapers slightly into robust cylindrical body that becomes laterally compressed on posterior quarter of trunk, and strongly compressed tail. Body scales are smooth, overlapping slightly, and occur in 19 to 23 (usually 21) rows at midbody; there are 192 to 211 narrowed ventral scales.

SIZE: Females average larger than males. Adult females range from 293 to 783 mm SVL (mean 605 mm); males from 339 to 654 mm SVL (542 mm). Males have longer tails than females, about 15% of body length (SVL) in males and ~11% in females.

OCCURRENCE: *L. frontalis* is a Vanuatu endemic with an occasional individual found in the Loyalty Islands. In Vanuatu, this species is known from Efate and Espiritu Santo; its occurrence has not been documented from the southern and more northern islands.

HABITAT CHOICE: This dwarf species is an inshore species, living primarily in sandy bottom habitats with rocky calcareous shorelines where it rests in cavities and beneath vegetation.

REPRODUCTION: *L. frontalis* is an egg-layer, but, as yet, its reproductive parameters and cycle have not been documented.

MISCELLANEA: Like its sympatric congener *L. colubrina*, *L. frontalis* has a diet of eels. It also shares the same resting sites and commonly occurs together in the same terrestrial aggregations.

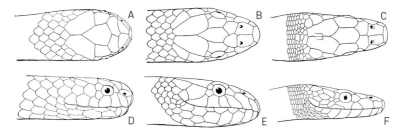

FIGURE 23. Head morphology of seasnakes and seakraits. Dorsal and lateral views of *Laticauda laticaudata* (A, D), *Hydrophis* (*Leioselasma*) *melanocephalus* (B, E), and *Pelamis platura* (C, F).

Laticauda laticaudata　　　　　　　　　　**Dark-lipped Seakrait**

Plate 31, Figure 23

APPEARANCE: Midsize to large seasnake with 25 to 79 black bands on a light to dark bluish-gray background. Few to many dark bands encircle body, and these bands are usually two to three times wider than the lighter interspaces. Ventral background is cream to yellowish white. Head has a bluish cap posteriorly and usually is light cream on top of head from eyes to snout; face, upper, and lower lips are dark. Tail is banded as is body and usually narrowly tipped in white. Head is robust and tapers slightly into a robust cylindrical body that becomes laterally compressed on posterior quarter of trunk and strongly compressed tail. Body scales are smooth, overlapping slightly, and occur in 19 to 21 rows at midbody; there are 219 to 252 narrowed ventral scales.

SIZE: Females and males may be nearly equal in size. Adult females range from 740 to 1,130 mm SVL, with tail length approximately 10% to 12% of SVL. In New Caledonia and probably also Vanuatu, mature females ranged from 885 to 1,295 mm SVL (mean 107 mm). The maximum reported total length is 1,360 mm.

OCCURRENCE: *L. laticaudata* is a resident of nearshore water in Fiji and Vanuatu; it is probably present in Tonga (although this requires confirmation) owing to its modest abundance around Niue. It is more abundant in the seas around New Caledonia. It has a broad distribution from eastern India and Southeast Asia throughout the waters of the Sundas and eastward to Taiwan and the Philippines.

HABITAT CHOICE: This seakrait occurs in a variety of clear tropical water habitats from bare muddy bottoms through seagrass beds to faces of coral reefs. Like other seakraits, it appears to forage mainly at night and then comes ashore to rest and digest in shaded sites within beach-side forest and beneath coral rubble.

REPRODUCTION: *L. laticaudata* is an egg-layer and deposits clutches of one to 14 eggs (usually seven or fewer).. The eggs are deposited in moist sites above the high tide line, usually deep within coral rubble mounds. This species breeds year-round in the Philippines; it is seasonal in New Caledonia, laying eggs from January to March.

MISCELLANEA: The Dark-lipped Seakrait co-occurs with the Yellow-lipped Seakrait. The latter is always more abundant than the former, which seldom exceeds 10% to 15% of the total local population of seakraits.

L. laticaudata is a fisheater, concentrating on anguilliform such as Moray Eels, although its diet includes other fishes.

Laticauda schistorhyncha　　　　　　　　　　**Central Pacific Seakrait**

Figure 22

APPEARANCE: Midsize seasnake with 18 to 31 black bands on medium-blue to grayish background. All dark bands encircle body, although uncommonly dorsal color is faded and bands disappear ventrally. Anterior two-thirds of head is dark; posteriorly it has yellowish band that encircles head and anterior neck with small medial projection toward the chin. Tail is banded as is body and usually narrowly tipped in yellow. Head is robust and tapers slightly into robust cylindrical body that becomes laterally compressed on posterior quarter of trunk and strongly compressed tail. Body scales are smooth, overlapping slightly, and occur in 23 rows at midbody; there are 187 to 195 narrowed ventral scales.

SIZE: Adult females average slightly larger than males. Adult females range from 498 to 806 mm SVL (628 mm); males from 465 to 750 mm SVL (612 mm). Tail length is 11% to 16% of SVL.

OCCURRENCE: *L. schistorhynchus* is a south-central Pacific endemic. It resides in the nearshore waters of Niue. It is commonly reported from Samoa and Tonga, although its presence there requires confirmation.

HABITAT CHOICE: *L. schistorhynchus* is a shallow water inhabitant and occurs in a variety of habitats inside the reef.

REPRODUCTION: Presumably, this small seakrait is an egg-layer. Its reproductive data have not been reported.

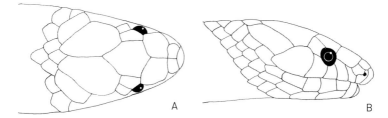

FIGURE 24. Head scalation of the Fiji Bolo (*Ogmodon vitianus*). Schematic drawing of dorsal (A), and lateral (B) views of the head.

Ogmodon vitianus   **Fiji Bola**

Plate 30, Figure 24

APPEARANCE: Small snake with uniform dark-brown or black body with iridescent sheen. Juveniles have broad, creamy-white parietal bar that shrinks with maturation and disappears in adults. Venter is white in juveniles and progressively darkens with growth, becoming most black with white patches in largest individuals; tail is black above and below in all individuals. Head is moderate sized and continuous with neck. Body is cylindrical, moderately robust, and ends in short, rapidly narrowing tail. Body scales are smooth, overlapping, and occur in 17 rows at midbody; there are 134 to 146 ventral scales.

SIZE: Adult females average larger than males. Females range from 208 to 315 mm SVL (mean 250 mm); males from 205 to 323 mm SVL (238 mm). Tail length is 16% to 17% of SVL. Hatchlings are about 110 to 115 mm SVL.

OCCURRENCE: This snake is known only from a few intermontane valleys in southeastern Viti Levu, Fiji.

HABITAT CHOICE: The Fiji Bola appears to be confined to relatively deep loam soils of forests or in gardens (particularly sweet potato patches) adjacent to or within forests.

REPRODUCTION: Sexual maturity is obtained in both sexes by 200 mm SVL. Clutch size is two to three eggs. No data are available on seasonality of reproduction.

MISCELLANEA: As a burrower, *O. vitianus* appears to be sensitive to dehydration and is found on the surface only after heavy rains have waterlogged the soil. This species preys on a variety of soft-bodied invertebrates, including insect larvae and earthworms.

Pelamis platura   **Yellow-bellied Seasnake**

Plate 31, Figures 22, 23

APPEARANCE: Midsize to large snake with dark-brown to black dorsum and bright-yellow, sometimes brownish, venter. Change from dark dorsum to light venter occurs sharply midlaterally, probably countershading, although some eastern Pacific populations are entirely yellow. Laterally compressed tail is barred or spotted in black to dark brown on yellow or cream background. Head is large, narrow ovate in dorsal outline, and tapers abruptly into cylindrical body that becomes increasingly compressed posteriorly. Ventral scales are small and range from 260 to over 400 scales. Body scales are small and abutting, in rows of approximately 50 to 70 scales encircling body. These scales are generally hexagonal and frequently tuberculate.

SIZE: The maximum reported total length is 1,130 mm; however, most individuals are less than 800 mm long. Individuals from Fijian and New Caledonian populations typically are less than 550 mm. Females are larger than males. Off the Pacific coast of Panama, females average 481 mm SVL (542 mm total length) and males 452 mm SVL (514 mm total length). Newborns are 220 to 260 mm total length.

OCCURRENCE: The Yellow-bellied Seasnake is an Indo-Pacific species. Its distribution extends from the east coast of Africa (from the Persian Gulf to the Cape of Good Hope) to the coast of Central America and the Galápagos Islands.

HABITAT CHOICE: Of all seasnakes, *P. platura* is truly pelagic. Often observed hundreds of kilometers from the nearest shoreline or reef flat, nevertheless, it is found most often in shallower waters over continental shelves or submarine plateaus, living in the upper levels of water column. *P. platura* regularly aggregates along ocean slicks.

REPRODUCTION: This seasnake is a livebearer and produces broods of two to eight neonates. They appear to be aseasonal breeders, with births year-round. The length of pregnancy is uncertain—probably five to seven months or slightly longer.

MISCELLANEA: The venom of *P. platura* is toxic to mammals; however, the toxicity is an adaptation that allows it to subdue its fish prey quickly and with minimum struggle that might injury the seasnake. These snakes eat a great variety of fishes and their larvae; all prey are captured in open water.

Homalopsidae: Mudsnakes

The homalopsids are a small group of rear-fanged aquatic snakes. There are about three dozen species occurring in freshwater and brackish habitats from India to southern China and southward through Sundaland to northern Australia and the Philippines. All species are strongly aquatic, rarely leaving the water. All share a set of aquatic adaptations that include valvular nostrils on the dorsal surface of the snout and small, dorsally directed eyes. Most species are freshwater inhabitants and feed primarily on fishes and frogs. The brackish water species also prey predominantly on fishes, although one species has become a decapod-crab specialist. Their venom is important because it helps them quickly subdue their prey. They bite the prey and continue to hold it with a chewing motion that delivers the venom deeply and quickly subdues it.

Cerberus rynchops   Dog-faced Mudsnake

Plate 30

APPEARANCE: Midsize to large snake with uniform gray ground color or dusky brown with darker brown transverse bands. Bands are incomplete and to not extend onto venter. Venter is grayish white with irregularly shaped dark blotches. Head is large, bluntly triangular, and distinct from neck; it is dorsoventrally flattened with protruding eyes on the dorsal surface. Body is robust and cylindrical with a thick and moderately long tail (16% to 28% of SVL). Body scales are strongly keeled (unicarinate), overlapping, and occur in 21 to 25 rows at midbody; ventral scales are large and range from 134 to 170 scales.

SIZE: Adult females appear to average larger than males. Females range from 560 to 900 mm SVL; males from 570 to 700 mm SVL. The maximum reported size is 1,270 mm total length. Newborns range from 130 to 160 mm SVL.

OCCURRENCE: This mudsnake is the only homalopsid in Oceania and inhabits Palau. Presently it is reported from Babeldaob, Ngerekebesang, and Oreor; however, it is likely to be widespread throughout the islands of Palau. Elsewhere, it occurs from coastal western India eastward across southern Southeast Asia to the Philippines and Lesser Sunda Islands.

HABITAT CHOICE: The Dog-faced Mudsnake lives in a variety of coastal habitats, predominantly shallow-water ones such as mangroves and associated mudflats and tidal creeks. It also inhabits more open waters of estuaries and the tidal reaches of rivers (into freshwater). It is a predominantly brackish and saltwater snake.

REPRODUCTION: *C. rynchops* is a livebearer. Females produce litters of five to 38 young. Depending upon the locality, reproduction is seasonal (predominantly December through April) or year-round. Evidence suggests that a female bears only one litter a year and that she does not become pregnant every year. Gestation also appears to vary geographically, and is perhaps affected by water temperature; the reported duration is from three to six months.

MISCELLANEA: *C. rynchops* preys mainly on fishes, particularly bottom dwellers such as gobies and catfish; shrimp can also be a main prey item.

The Dog-faced Mudsnake is a venomous, rear-fanged snake. There are no reports of human fatalities.

The preceding information applied to the pre–September 2012 concept of *C. rynchops*. Now *Cerberus* consists of five species, one of which (*C. dunsoni*) is endemic to Palau and uniquely characterized by 23 scale rows at midbody. *C. dunsoni* ranges from 228 to 687 mm SVL; tail length is 21% to 26%.

TESTUDINES: TURTLES

Everyone recognizes a turtle. The body encased in a shell is a dead give-away. No other group of extant vertebrates has its vertebrae and ribs fused into a dorsal shield (carapace) and part of its sternum fused with dermal bones forming a ventral shield (plastron). Both the pectoral and pelvic girdles lie within the carapace and plastron. Additionally, all turtles have a neck of eight vertebrae. This constant neck morphology has internal differences that result in two different mechanisms of head retraction. While most turtles withdraw their heads directly backwards with a vertical flexure of the neck, and are awkwardly called S-necked turtles (Cryptodira), sidenecks (Pleurodira) fold their necks to the side through a horizontal flexure of the neck.

These two groups—pleurodires and cryptodires—represent two ancient lineages that diverged at least 150 million years ago in the Jurassic. Cryptodires are the most widespread and diverse of living turtles; they include the freshwater turtles, principally of the Northern Hemisphere; tortoises; and seaturtles. Today, pleurodires are strictly freshwater turtles and occur only in the Southern Hemisphere, with their greatest diversity in Australia and South America. A sideneck on a tropical Pacific island is an intentional or accidental introduction by humans. All reports of sidenecks on Pacific islands are less than 100 years old.

Chelidae: Sideneck Turtles

There are three families of sideneck turtles (Pleurodira). The Chelidae have generic representatives in South American and Australia-New Guinea; the two other families contain African and South American species. Chelids differ from the other sidenecks by the presence of a cervical scute on the front edge of carapace and by the absence of a mesoplastral bone in the plastron. All sidenecks currently reported from Pacific islands are chelids and derive from Australia or New Guinea populations.

Chelodina longicollis   **Eastern Long-necked Turtle**

Plate 33

Chelodina turtles are easily recognized by their extremely long, flexible necks. The neck is as long as the carapace or nearly so and bends in a serpentine fashion as the turtle searches for prey.

APPEARANCE: Midsize freshwater turtle with dark-brown, elliptical carapace and large, elliptical plastron that nearly cover entire underside of carapace. Plastron is light yellow with plastral sutures edged in dark brown. Carapace is smooth and depressed medially; second through fourth vertebral scutes are grooved medially. Long, snakelike neck and bullet-shaped head are unique among Pacific island turtles, as is pebbly texture of neck skin.

SIZE: Females average slightly larger than males. Average and maximum carapace lengths vary geographically. Females attain a maximum 28 cm SCL and males 25 cm SCL. Most adults are less than 20 cm SCL. Hatchlings are ~30 mm SCL.

OCCURRENCE: *Chelodina longicollis* is a recent arrival in Rarotonga, Cook Islands. Introduction and establishment dates are uncertain, although this sideneck appeared to have an established breeding population by 2009. *C. longicollis* is widespread in eastern Australia, found in coastal drainages from South Australia to central Queensland and in the Murray-Darling drainages. A chelid turtle, perhaps *Chelodina*, has been reported from Palau, but this report remains unconfirmed.

HABITAT CHOICE: This long-necked turtle is principally a bottom-walker that forages in the slow-moving portions of streams. It is is a common resident of ponds and marshes.

REPRODUCTION: In Australia, this sideneck is a midsummer nester (November and December). It nests at night, high on the grassy and sandy banks of its resident streams. A single clutch of six to 24 hard-shelled eggs is laid each year. Incubation is moderately long, ranging from 18 to 28 weeks.

MISCELLANEA: Seemingly a nonselective carnivore and scavenger, *C. longicollis* consumes a wide variety of aquatic invertebrate and vertebrate prey. When disturbed, its major defense is well-developed musk glands, opening on the plastral bridge, that release a foul-smelling liquid musk. This turtle is ecologically and physiologically tolerant and can survive in ephemeral ponds by estivation, by burrowing into the bottom, or by migrating overland to another body of water.

Cheloniidae: Hard-shelled Seaturtles

Cheloniidae includes all six species of hard-shelled seaturtles. These seaturtles have well-developed bony shells covered with large keratinous plates (scutes), similar to those of most other turtles. Shell and protruding body parts are hydrodynamically streamlined, shell dorsoventrally thickest anteriorly and tapering rearward. Forelimbs are flipper-like, with no free digits. The hand portion forms more than two-thirds of the flipper surface. The forelimbs are the main locomotor source, and their figure-8 movement cycle mimics the flight movement of a bird's wing. Seaturtles literally fly underwater with the up-and-down strokes of both flippers propelling the turtle forward.

Seaturtles seldom emerge from the sea for any reason other than reproductive nesting by females, although an exception is seen in the Hawaiian and Galápagos Green Seaturtles, which often bask on beaches. Nesting occurs typically in multiyear cycles. During the nesting year, each female emerges multiple times to dig a nest in the sand above the reach of high tide and deposit a large clutch of eggs, although the number of eggs in each clutch declines with each successive nesting. Both the interseasonal frequency of nesting and the number of egg-laying sessions within the nesting season are thought to be driven by nutrition. Abundant and high-quality foods allow females to develop fat stores quickly and obtain the energy reserves necessary to produce viable eggs and to move to their nesting beaches and return to their feeding grounds when all eggs are laid.

Caretta caretta   **Loggerhead Seaturtle**

Plate 32, Figure 25

Loggerheads are uncommon in nearshore waters of tropical Pacific islands. Adult loggerheads are nearshore residents over the continental shelves of tropical to warm temperate seas.

APPEARANCE: Very large seaturtle with reddish-brown, obovate carapace and light-yellow elliptical plastron. Reddish-brown to yellowish head is distinctly large and bears two pairs of square scales (prefrontals) dorsally on snout. Carapace has five vertebral scutes, medially keeled in younger individuals, and usually five pairs of pleural scutes. First pleural scute on each side is proportionately small and touches outer rear edge of cervical scute. On each side, plastral bridge of adults bears three inframarginal scutes without pores.

SIZE: Adults might be sexually dimorphic; however, sufficient data for adult males are lacking. There are insufficient data available for the sizes of nesting females among the islands covered in this guide. Nesting females in Australia average 100 cm SCL (70–146 cm). Pacific hatchling loggerheads average 43 mm SCL (39–50 mm).

OCCURRENCE: Loggerheads occur irregularly among the tropical Pacific islands. They reportedly nested infrequently in Tokelau in the early 1980s; the status of that nesting population is currently unknown. Some nesting occurs in the Isle of Pines, New Caledonia; scattered nesting probably occurs in Vanuatu.

HABITAT CHOICE: As adults, Loggerheads are nearshore turtles occurring predominantly in the shallower waters (30 m or less) over continental shelves. Hatchlings and juveniles, however, are oceanic residents and spend several years in the oceanic gyres, often on the opposite side of the ocean from their hatching beach and their future feeding areas.

REPRODUCTION: The nesting season for Australian loggerheads extends from October through March; presumably nesting in Vanuatu occurs during this interval as well. Females are solitary nesters on sandy beaches. Nesting occurs above the high-tide line, typically in vegetation-free areas. Clutches of white, flexible-shelled spherical eggs are large, with between 80 and 190 eggs, averaging about 125 eggs. Incubation is seven to 11 weeks.

MISCELLANEA: Loggerheads commonly nest on subtropical and warm temperate beaches, such as those along the coast of eastern Australia and in the southeastern United States. Presumably the tropical southwestern Pacific populations are also migratory and the nesting females derive from areas distant from their nesting beaches.

Loggerheads are carnivorous, generally preying on larger species of mollusk and decapod crustaceans.

Chelonia mydas   **Green Seaturtle**

Plate 32, Figure 25

Green Seaturtles are the most common marine turtle of the tropical Pacific. At one time, they nested on most sandy beaches and lived in reef flats of the largest islands to the smallest atolls.

APPEARANCE: Large to very large seaturtle with bright, broad elliptical carapace and light-yellow to white plastron. Each carapacial scute bears pattern of dark, radiating lines and spots on lighter background. Dark-brown head is small with single pair of elongate rectangular scales (prefrontals) dorsally on snout. Carapace has five large, smooth vertebral scutes and four pairs of pleural scutes. First pleural scute does not touch cervical scute. On each side, plastral bridge bears four inframarginal scutes without pores.

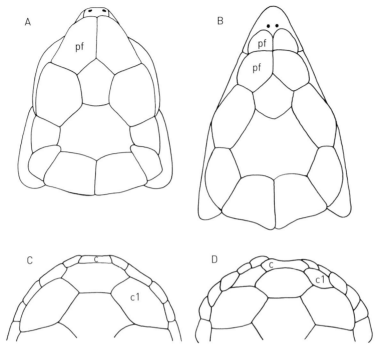

FIGURE 25. Features of the head and shell useful for the identification of hard-shelled seaturtles (Cheloniidae). Schematic of dorsal head scales of *Chelonia mydas* (A) and *Eretmochelys imbricata* (B), and of the anterior margin of the carapace of *C. mydas* (C) and *Caretta caretta* (D). *C. mydas* has a pair of prefrontal scales (pf); all other cheloniid seaturtles have four prefrontal scales. The cervical scute (c) of *C. mydas* and *E. imbricata* does not touch the first costal scute (c1); whereas the cervical scute touches the first costal scute in *Lepidochelys olivacea* and *C. caretta.*

SIZE: Adults are sexually dimorphic, with males averaging a few centimeters (SCL) shorter than females. Nesting females vary in size among different Pacific populations. Galápagos females are the smallest and average about 82 cm (68–106 cm); Hawaiian females average about 92 cm (80–106 cm); and New Caledonian and Australian females, about 106 to 108 cm SCL, respectively. Pacific hatchling *C. mydas* average between 46 and 50 mm SCL (range is 40–52 mm). Like adults, there are differences among populations.

OCCURRENCE: The Green Seaturtle occurs broadly in all tropical seas: Atlantic, Indian, and Pacific. It probably occurred in all island groups of the preceding checklists except Clipperton Island. Current absence from an island's list results either from no recent report confirming its occurrence or local extirpation.

HABITAT CHOICE: Green Seaturtles are nearshore turtles as large juveniles and adults, occurring predominantly in the vicinity of coral reefs and reef flats. Hatchlings and small juveniles are pelagic inhabitants.

REPRODUCTION: Some Pacific island populations nest year-round, although the peak nesting season varies. Green Seaturtles nest year-round in the Society Islands with peak nesting between September and February. The Hawaiian population nests only on the French Frigate Shoals between May and September; and most nesting in this population occurs in June and July. Females are solitary nesters, but often more than one female will be present on a nesting beach on most evenings if the population level is high. Females prefer wide sandy beaches with no or low high-tide berms, and dig their nests in vegetation-free areas above the high-tide zone. The clutch size of white, flexible-shelled spherical eggs is variable, ranging between 40 and 150 eggs (the average is 100 eggs or fewer) in most Pacific populations. Incubation is six to nine weeks.

MISCELLANEA: While living in nearshore habitats with apparently appropriate nesting beaches nearby, most females migrate to distant nesting beaches. All Hawaiian females nest in the French Frigate Shoals area. Some central Pacific females nest on islands of the Great Barrier Reef, having migrated hundreds of kilometers from Samoa and Tonga.

Green Seaturtles are herbivores as large juveniles and adults. In most Pacific populations, the turtles are dependent upon a variety of algae. If seagrasses are available, they are consumed.

Eretmochelys imbricata   **Hawksbill Seaturtle**

Plate 32, Figure 25

Hawksbills are less abundant among tropical Pacific islands than Green Seaturtles. They are, nonetheless, commonly seen in the flats and seaside faces of healthy coral reefs.

APPEARANCE: Large seaturtle with brightest carapace; plastron is light yellow to near white. Carapacial scutes appear polished with bold, mottled pattern of dark spots and lines. Carapacial scutes are large and distinctly overlap one another on their posterior edges. Sides of carapace are serrated strongly on posterior third of carapace. Dark-brown head is moderately small and elongate; two pairs of pentagonal scales (prefrontals) occur dorsally on snout. Carapace has five large vertebral scutes, each keeled, and four pairs of pleural scutes. First pleural scute on each side does not touch cervical scute. On each side, plastral bridge bears four inframarginal scutes without pores.

SIZE: Adults might be sexually dimorphic, with males averaging a few centimeters (SCL) smaller than females. A large sample of females nesting in the Solomon Islands averaged 80.5 cm SCL (68–93 cm). Nesting females from the central Pacific appear to be slightly smaller. Pacific hatchling Hawksbills average 39 mm SCL (38–41 mm).

OCCURRENCE: Hawksbills occur in all tropical seas. They probably occurred in all Pacific island groups except Clipperton and Malpelo. Current absence from an island's list is the result of either no recent report confirming occurrence or local extirpation.

HABITAT CHOICE: Hawksbills are nearshore turtles, occurring predominantly in the vicinity of coral reefs.

REPRODUCTION: The principal, perhaps exclusive, nesting season for the central Pacific Hawksbill extends from September to February. Females are solitary nesters, typically emerging on rock-free sandy beaches with no or low high-tide berms. They dig their nests under vegetation, including in beachside forest and scrub. Clutches of white, flexible-shelled spherical eggs are large, with between 70 and 220 eggs, averaging about 130 eggs. Incubation is seven to nine weeks.

MISCELLANEA: The Hawksbill diet is broad (the species is opportunistic omnivorous) but consists primarily of sponges. This is unusual, because few vertebrates can handle a diet primarily of sponges because of the high silica content of sponge tissue and the toxicity of many sponges. Hawksbills also prey on other invertebrates including mollusks, crustaceans, jellyfish, sea urchins, and other bottom- and reef-dwelling animals.

Hawksbills have experienced high human predation owing to the beauty of their carapacial scutes. These scutes become the tortoise-shell jewelry much admired by many different cultures. Although Hawksbills are legally protected in most Pacific nations, they are still harvested for their shells in some areas.

Lepidochelys olivacea   **Olive Ridley Seaturtle**

Plate 32

Olive Ridleys are pelagic turtles, common in the eastern Pacific and uncommonly seen among central and western Pacific islands. They are the smallest of the Pacific seaturtles.

APPEARANCE: Large seaturtle with dark-olive, cuneate carapace and light-colored, elliptical plastron. Light- to dark-olive head is large and bears single pair of short rectangular scales (prefrontals) dorsally on snout. Carapace has five or more vertebral scutes, each with distinct medial keel that disappears in older individuals. Number of pleural scutes on each side is also variable, ranging from five to nine; number on left and right sides often differ. First pleural scute on each side touches outer rear edge of cervical scute. On each side, plastral bridge bears four inframarginal scutes, each with single pore on its anterior outer margin.

SIZE: Adults show a slight sexual dimorphism, with males averaging a few centimeters (SCL) larger than females. Females nesting in Central America averaged about 66 cm SCL (58–75 cm). Hatchlings average ~40 mm SCL (35–45 mm).

OCCURRENCE: Olive Ridleys are vagrants among the tropical Pacific islands. Although they have a pantropical distribution, Olive Ridleys are seen regularly only in the vicinity of their nesting beaches; the major ones are on the west coast of Central America from mid-Mexico to Costa Rica, the Surinam coast of northeastern South America, the northeastern coast of India, and also occur spottily along the west coast of Africa.

HABITAT CHOICE: Olive Ridleys are considered nearshore inhabitants, feeding and migrating in the shallow waters over the continental coast.

REPRODUCTION: These turtles do not nest on Pacific islands. They nest only on continental coasts and in mass nestings called "arribadas," where hundreds of females emerge on the same stretch of beach beginning in late afternoon and continuing to arrive through early evening. Clutches of white, flexible-shelled spherical eggs are large, with between 70 and 220 eggs, averaging about 130 eggs. Incubation is seven to nine weeks.

MISCELLANEA: Olive Ridleys are often mentioned as being the most abundant seaturtle. This assessment may be correct, but it is driven by the aggregation of females in large "flotillas" prior to nesting, then the mass and near simultaneous emergence of these females on nesting beaches.

The Olive Ridley is a catholic carnivore. It preys on a variety of invertebrates from jellyfish to salps, an assortment of mollusks and crustaceans, and fishes. The species captures its prey throughout the water column from the surface to the bottom.

Dermochelyidae: Leatherback Seaturtles

Dermochelyidae is a family with a single extant species, *Dermochelys coriacea*, which occurs worldwide in temperate and tropical seas. This seaturtle derives its name from a tough leathery skin overlying a greatly reduced dorsal (carapace) and ventral (plastron) bony shell. Most skeletal components of the typical turtle shell are present, except there are no peripheral bones on the margin of the shell, and a mosaic of small bone plates covers the carapace and plastron skeleton. Skin and shell modifications are apparently adaptations for a pelagic life of extensive swimming and diving. The modifications have resulted in a hydrodynamic body plan. The flattened fusiform shape and locomotion powered by large flipper-like forelimbs yield swimming speeds clocked to nearly 30 knots. Like penguins and the hard-shelled seaturtles, leatherbacks are aquatic flyers; they swim with their forelimbs. The forelimbs move through a figure-8 stroke that powers forward motion with both the up and the down strokes. The large webbed hindfeet serve as rudders at most swimming speeds and are also used for locomotion at low speeds.

Leatherbacks travel long distances among oceanic and coastal feeding areas and to tropical beaches during their multiple-year reproductive cycle. These movements probably place them near all Pacific islands during any individual's lifetime, but usually in deeper offshore waters, hence they are uncommonly seen and reported by local fishermen or marine biologists. Their large body size and limbs adapted for swimming handicap the females' nesting. Their egg-laying beaches are widely scattered, with none among the islands of Oceania.

Dermochelys coriacea   **Leatherback Seaturtle**

Plate 32, Figure 26

Leatherbacks are the largest extant turtles and probably also the species in which adults annually travel the greatest distances. Atlantic leatherbacks feeding in early summer in Canadian waters and the Bay of Biscay come from the waters of northeastern South America, to which they return in fall and winter. In the Pacific, females nesting in Central America head south into oceanic waters south of the Galápagos Islands; of the New Guinea nesting populations, some will cross the entire Pacific to the coast of California and northern Mexico; others head south into the Indian Ocean east of Australia.

FIGURE 26. Head morphology of *Dermochelys coriacea*. The front of the upper jaw of leatherbacks has large curved tusk-like projections of the keratinous tomium or labial sheath. This feature occurs in no other seaturtle.

APPEARANCE: Giant seaturtle with black, obovate carapace and black-and-white elliptical plastron. Carapace is regularly flecked with white or pink, often concentrated on crest of five longitudinal ridges running full length of carapace; plastron is similarly ridged. Head, neck, body, limbs, and tail are also black flecked with white. Large head possesses deeply notched upper jaw bordered anteromedially on each side by large denticulate cusp. Carapace is leathery, lacking horny scutes of cheloniid seaturtles; horny scutes are briefly present in hatchlings.

SIZE: Adults do not appear sexually dimorphic. Adults of the Pacific, Indian, and Atlantic populations have different average sizes: Pacific adults average smaller. Females nesting on the Pacific coast of Central America averaged about 140 cm SCL (125–160 cm). Hatchlings average ~58 mm SCL (50–68 mm).

OCCURRENCE: Leatherbacks are infrequently seen among tropical Pacific islands, probably because their prey does not concentrate in large masses in tropical seas. They have the largest geographic distribution of the seaturtles occurring in northern and southern cool to temperate waters of both the Pacific and Atlantic Oceans.

HABITAT CHOICE: Leatherbacks are pelagic, feeding and migrating in the open ocean over both the continental shelves and abyssal plains. In some places—such as California, eastern Canada, and Atlantic coastal France—they occur seasonally in coastal waters where jellyfish and other sea jellies concentrate.

REPRODUCTION: These turtles nest on continental coasts, only on broad sandy beaches that lie near deep nearshore water. Clutches of white, flexible-shelled spherical eggs are small in number compared with those of other seaturtles: they have between 30 and 166 eggs, averaging about 77 eggs. However, the individual eggs are larger than those of other seaturtles. A nest can be nearly a meter deep. Incubation requires seven to eleven weeks, depending on sand temperature around the egg clutch.

MISCELLANEA: Leatherbacks are the only turtle capable of maintaining body temperature above the ambient temperature in cool to cold water. This ability derives from several morphological and physiological adaptations. These adaptations include the high oil content of the skin, which serves as insulation in the same manner as whale blubber, and a counter-current circulatory network in the limbs that reduces heat loss from the bloodstream.

Leatherbacks are dietary specialists as adults. They prey predominantly on jellyfish, pelagic tunicates, salps, and other sea jellies. They are active divers and make some of deepest dives of any vertebrate, apparently in search of thermoclines where their prey concentrate.

Chelydridae: Snapping Turtles

Snapping Turtles are aquatic turtles of the Americas, principally North America. Their large heads and open-mouthed, lighting-strike lunges when molested have given them their name. There are two genera: *Chelydra* and *Macrochelys* (the Snapping Turtle and the Alligator Snapping Turtle, respectively). The former occurs broadly in North America into Latin America; the latter is confined to the rivers of the Gulf Coast of the United States. Both taxa have large heads and sharp, beaked jaws, a flattened but distinctly longitudinally keeled carapace, a reduced plastron, and a long tail with a double row of large upright scales on its upper outer edges. They are carnivorous.

Chelydra serpentina   **Snapping Turtle**

Plate 33, Figure 27

Once seen, this turtle—with its rugged shell and aggressive defensive behavior—is not readily forgotten.

APPEARANCE: Brown, ridged carapace is roughly square in outline with rounded fore edges and strongly serrated posterior margin. Carapace readily covers body and limbs; light-colored plastron is greatly reduced to cross-shaped shield. This reduction allows Snapping Turtles to avoid sprawled gait; instead they bring their limbs beneath them and walk with body held high above ground. Head is large and slightly flattened; mouth is large and keratinous jaw sheath is sharp-edged. Skin is dull brown and covered with numerous folds, papillae, and tubercles.

Unlike most other turtles, tail is long, nearly length of carapace, and has double row of erect plates on its dorsolateral margins.

SIZE: The maximum reported length is 494 mm SCL for a male and 366 mm SCL for a female. Size varies geographically; males always average larger. Adult females range from 185 to 366 mm SCL; males from 180 mm to 494 mm SCL. Hatchlings are 23 to 33 mm SCL.

OCCURRENCE: *C. serpentina* is an eastern North American turtle, ranging east of the Rocky Mountains from southern Canada to the Gulf Coast and southern Florida. Its presence in Guam is recent, first reported in 1998, and seemingly confined to southern Guam. Infrequent reports of hatchlings or small Snapping Turtles suggest, however, that this turtle has become established on Guam.

HABITAT CHOICE: Snapping turtles are aquatic generalists and live in a wide variety of freshwater and brackish habitats, from large rivers to small creeks and from lakes to ponds, marshes, and swamps. They are prefer soft-bottom habitats with vegetation.

REPRODUCTION: In southern Florida, reproduction may occur year round; however, in most of the snapper's range, egg-laying occurs in midspring. Clutch size also has a geographic component, with larger clutches laid in the north. Most clutches range between 25 and 45 eggs. Incubation is typically 75 to 95 days, shorter in hotter nests and slower in colder nests, with a maximum of 125 days.

MISCELLANEA: Snapping Turtles are omnivores with a carnivorous preference. Plant material is a regular component of a snapper's diet, as are fishes and invertebrates, including insects. Any aquatic vertebrate is acceptable prey if it can be captured. Snappers also feed on carrion.

Emydidae and Geoemydidae: Pondturtles

The Emydidae and Geoemydidae are the pondturtles of the Western and Eastern Hemispheres, respectively, but not exclusively. A single emydid species—the European Pondturtle—occurs from northern Africa and western Europe to western Russia and Ukraine. A single geoemydid genus—the Neotropical Woodturtles, which consists of nine species—occurs from southern Mexico into northern South America. These two families share many morphological traits and are differentiated only by internal ones. All have firm, bony shells covered with horny scutes and small heads. Most are small to medium-sized turtles. None of the pondturtles are native to Oceania.

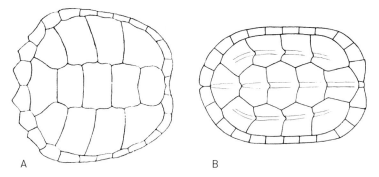

FIGURE 27. Shell shape in freshwater turtles. Dorsal outlines of the carapace in (A) *Chelydra serpentina* and (B) *Mauremys reevesii*. The broad, flattened shell with a strongly serrated posterior margin is a distinctive feature of *Chelydra*.

Trachemys scripta   Red-eared Slider

Plate 33, Figure 28

APPEARANCE: Small to midsize freshwater turtle with brightly marked, oval to oblong carapace and light-colored plastron, often with large, dark plastral blotches. Background of carapace is dark brown to nearly black and is emblazoned with numerous yellow stripes on each vertebral and pleural scute. Bright carapacial pattern can be lost with age. The English name derives from brightly striped head; most of the stripes are black-bordered yellow ones except for the bright-red postorbital stripe that broadens over ear.

SIZE: Adults are sexually dimorphic; males average smaller than females. Generally, adult males range from 100 to 200 mm SCL (average ~160 mm); females from 150 to 250 mm SCL (~200 mm). Adults in southern populations appear to average slightly smaller than those of northern populations. As yet, there is no comparison of adult sizes among and within the numerous invasive populations. Hatchlings average ~36 mm SCL (range, 25–44 mm).

OCCURRENCE: *T. scripta* occurs widely throughout eastern North America. The original Red-eared population, *T. s. elegans*, was mainly resident of the Mississippi drainage system and westward through Texas. The Pacific invasive populations are widespread, occurring now in Guam and the southern Marianas, the Bonin Islands, the Hawaiian Islands (Kauaʻi, Maui, and Oʻahu), and French Polynesia.

HABITAT CHOICE: The Red-eared Slider is a habitat generalist and occupies most aquatic habitats. However, it prefers slow-moving waters of ponds, lakes, and reservoirs, especially shallow water (1–2 m deep) with soft bottoms and thick vegetation.

REPRODUCTION: This species nests in spring to early summer in its native range. Being very adaptable, it probably adjusts its reproductive cycle to the local climate. Local breeding cycles are not as yet well studied in Pacific populations, and probably most island populations have a broad nesting season. The Red-eared Slider lays moderate-sized clutches with an average of about 10 but as many as 30 white, flexible-shelled spherical eggs; clutch size is correlated positively with body size, with bigger females laying more eggs. Incubation is typically seven to nine weeks.

MISCELLANEA: The Red-eared Slider is one of three North American subspecies or geographic morphs of *Trachemys scripta*. In the mid-20th century, *T. s. elegans* became a popular and durable pet in North America. Hatchlings were widely sold in "five-and-dime" variety stores and became known as the "dime-store turtle." Its popularity was so great that a commercial turtle-raising industry developed because harvesting wild hatchlings could not meet the demand. This industry was concentrated in the southern part of the Mississippi Valley, which also was the center of the commercial poultry industry. Offal from the latter industry was recycled as food for the turtle industry. Because fowl are a good source of salmonella, the pet turtles became carriers. In the 1960s, a large number of US childhood salmonella infections were traced to the dime-store turtles. Federal public health services banned the selling of hatchling turtles in the United States, and the turtle-raising industry shifted its sales to the export market. Red-ear Sliders then became a popular, inexpensive pet purchase throughout Europe, Asia, and Australia.

When pet owners are diligent, their pet turtle remains healthy and soon outgrows the confines of its aquarium. For many, owners solved this problem by releasing their pets into a local lake or stream. Many of these released turtles survived, and now Red-eared Slider turtle populations are well established throughout areas of earlier commercial distribution.

Mauremys reevesii   **Chinese Three-keeled Pondturtle**
Plate 33, Figures 27, 28
APPEARANCE: Small to midsize freshwater turtle with light- to dark-brown, oval to oblong carapace and yellow plastron bearing brown blotches on

FIGURE 28. Lateral head striping in Pacific pond turtles. (A) *Mauremys reevesii.*
(B) *Mauremys sinensis.* (C) *Trachemys scripta elegans.*

bridge and each plastral scute. Carapace has strong middorsal (vertebral) ridges bordered on each side by lower longitudinal ridge on upper ends of pleural scutes. Head and neck are yellow striped laterally, usually with yellow flattened C-stripe behind eye; tympanum often edged in yellow; continuous dorsolateral stripe from temporal area onto neck.

SIZE: Size varies geographically and is sexually dimorphic in all populations. Adult females are larger than males; in general, females average about 154 mm SCL (range, 121–217 mm), males 145 mm SCL (121–236 mm). In one Japanese population, females averaged ~190 mm (maximum 237 mm SCL) and males ~150 mm (max. 185 mm). The maximum reported SCL barely exceeded 300 mm. Hatchlings range from 25 to 37 mm SCL.

OCCURRENCE: The presence of *M. reevesii* in Palau is questionable and based on a photograph that appeared in a Palauan-produced guidebook. No locality was associated with the photograph. In Guam, a Chinese Three-keeled Pondturtle belonging to a herp-hobbyist was reported to have been collected locally in the early 2000s; no subsequent specimens have been found. The native range of *M. reevesii* is southeastern China, the southern half of the Korean peninsula, and Taiwan. It is also widespread in the southern islands of Japan; however, these populations appear to derive from a late-18th-century introduction from Korea.

HABITAT CHOICE: The Chinese Three-keeled Pondturtle is a creature of slow or still waters and is typically found in marshes, ponds, and canals, all with soft bottoms and abundant vegetation. It is strongly aquatic and seldom leaves the water other than to lay eggs.

REPRODUCTION: Egg laying occurs in June and July in China. In a Japanese population, females can lay as many as three clutches in a season, although one or two clutches is the norm. Clutches contain four to nine oblong, flexible-shelled eggs. Incubation commonly ranges from six to 10 weeks; in colder areas, hatchlings overwinter in the nest.

MISCELLANEA: These pondturtles are omnivores that eat a variety of aquatic vegetation and small invertebrates and vertebrates.

Mauremys sinensis   **Chinese Stripe-necked Pondturtle**
Plate 33, Figure 28

APPEARANCE: Small to midsize freshwater turtle with reddish-brown to nearly black oblong carapace and slightly smaller, yellow plastron with dark-brown blotches in each scute. Carapace is slightly depressed, and three discontinuous longitudinal keels of juveniles disappear in subadults and adults. Head and neck have olive ground color and are intensely lined with numerous narrow, longitudinal yellow to greenish-yellow stripes; each stripe is highlighted by dark borders. Limbs are similarly striped in yellow on olive background.

SIZE: Size varies geographically and is sexually dimorphic in all populations. Adult females are larger than males, averaging 192 mm SCL (range, 114–271 mm); males 140 mm SCL (107–200 mm). Hatchlings average ~35 mm SCL.

OCCURRENCE: The establishment of a breeding population of *M. sinensis* in Guam is unconfirmed. The report of this species derives from several *M. sinensis* in the possession of a local reptile-hobbyist, who purportedly found them in the Agana Springs area. Its native distribution is coastal southern China from Shanghai southward to northern Vietnam and possibly eastern Laos. It also occurs in Hainan and Taiwan.

HABITAT CHOICE: *M. sinensis* prefers slow-moving and stationary waters of sluggish streams, canals, ponds, lakes, and marshes. It appears to select locations with emergent vegetation.

REPRODUCTION: In Taiwan, females nest from mid-March into early May. Females produce one to three clutches each year in a Japanese population. Clutches contain seven to 17 eggs (average 12). Incubation is probably six to 10 weeks and depends on nest temperature.

MISCELLANEA: Juveniles and adult males are largely carnivorous, eating a wide variety of invertebrates, although seemingly excluding snails. Young females share this diet, but they become increasing herbivorous, although not exclusively so, as they mature.

Testudinidae: Tortoises

Tortoises are a successful group of turtles and include about 60 species. They occur naturally on all continents except Australia, and are immediately recognizable by their columnar or elephantine hindlimbs. All tortoises are terrestrial; they occupy a full range of land habitats from deserts to tropical rainforests. They range in size from the South African Speckled Padloper, with a maximum carapace length (CL) of only 10 cm, to the insular giants of Aldabra and the Galápagos Islands, with shell lengths that range from 50 cm to more than 150 cm CL.

All Neotropical tortoises are members of the genus *Chelonoidis*. Galápagos tortoises derive from South American ancestors, which reached the Galápagos millions of years ago, probably by floating like bobbing corks on westward-flowing ocean currents from the mainland. Until recently, all tortoises of the Galápagos were considered members of a single species, although each island had one or more distinct shell forms and each was recognized as a distinct race or subspecies. That interpretation changed when a conservation genetics laboratory at Yale University demonstrated that most named races were genetically different from one another, and each represented a unique lineage or species. The following list of species relies on their molecular studies.

As many as 15 species of giant tortoises used to live in the Galápagos Islands. Today, 11 species survive, although one species (*C. abingdoni*) was represented only by a single individual—Lonesome George—until his death in the summer of 2012. Several other species persist only as a result of a captive breeding program in the Galápagos. The survival of all these species, even the most abundant ones, depends on aggressive conservation efforts to protect them from alien predators such as ants, rats, cats, dogs, and pigs, as well as their habitats' destruction by goats, donkeys, and humans.

Galápagos tortoise species are recognized by shell (carapace) shape, which is domed rather than saddleback. This characterization is discerned easily at its extremes in the older adult tortoise, but it is much less evident in younger adults and not at all in juveniles. Further, a number of species have shell shapes that are intermediate between the two extremes. Van Denburgh, in his classic study of Galápagos tortoises, developed a complex formula of proportional shell measurements to differentiate between species of the two shell shapes. These formulae work for most adults; however, few readers will have the time or opportunity to record nine measurements and calculate the formula

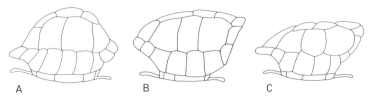

A   B   C

FIGURE 29. Examples of Galápagos tortoise shell shapes. *Chelonoidis vanden-burghi* (A) has a domed shell, *C. darwini* (B) an intermediate carapace, and *C. hoodensis* (C) a saddleback carapace. These three shells schematically display the gradation from domed to saddleback-shaped shells. There are numerous intermediate shell shapes.

for each individual seen. Instead, the set of shell outlines above will assist identification, although identification of wild tortoises is best done by relying on location, because each species occurs separately.

Chelonoidis abingdonii   **Pinta Giant Tortoise (Lonesome George)**
Not illustrated

APPEARANCE: Large to very large, distinctly saddleback tortoise with an upside-down goblet-shaped anterior opening. Posterior margin of shell is slightly flared, and bridge connecting carapace to plastron is moderately broad, about 44% to 46% of carapace length. Head is small and blunt, with weakly bicuspid upper jaw. Shell is black above and below; skin of head, neck, and limbs is dark gray. In males, head is often strongly suffused with yellow.

SIZE: The available data indicate that adult size ranged from 44 to 92 cm straight carapace length (SCL). The last surviving male was about 100 cm SCL.

OCCURRENCE: The original population on Isla Pinta was decimated by 19th-century whalers. A single surviving male discovered in 1971 was removed to the Charles Darwin Research Station on Santa Cruz in 1972 for safety and potential breeding.

HABITAT CHOICE: Pinta is a dry, rugged island with large expanses of bare rock and lava flows that separate low scrub forest and patches of Opuntia (prickly-pear) cactus.

REPRODUCTION: No data are available; the species is now extinct.

MISCELLANEA: A single *C. abingtoni* survived until the summer of 2012. He was known as "Lonesome George" and had become a conservation icon with the remote hope that he would breed and save his species.

Chelonoidis becki   Wolf Giant Tortoise

Not illustrated

APPEARANCE: Large to very large tortoise with saddleback shell. Rim of anterior shell opening is distinctly upturned; posterior margin of shell is nearly vertical. Bridge connecting carapace to plastron is moderately narrow, about 37% to 41% of carapace length. Head is moderate-sized, blunt, with weakly bicuspid upper jaw. Shell is gray above and below; skin of head, neck, and limbs is similarly gray. Lower jaw and throat of males are commonly faded yellow.

SIZE: Size data are limited. Males range from 46 to 105 cm SCL; females from 55 to 85 cm CCL (curved carapace length).

OCCURRENCE: This tortoise lives on the northern and western slopes of Volcan Wolf, Isla Isabela.

HABITAT CHOICE: The slopes of Volcan Wolf are steep and rugged and covered by dense thickets of thorny shrubs and small trees interspersed with clumps of grass.

REPRODUCTION: Data on reproduction are not accessible.

MISCELLANEA: The difficulty of human access to its area offers a level of protection to this species. Population size is estimated at approximately 1,500–2,000 individuals.

Chelonoidis chathamensis   San Cristóbal Giant Tortoise

Not illustrated

APPEARANCE: Large tortoise with shell shape intermediate between domed and saddleback. Typically, females are more domed and mature males develop distinct saddleback appearance. Rim of anterior shell opening is slightly upturned in saddleback shell and smoothly curved in domed condition; posterior margin of shell is nearly vertical. Bridge connecting carapace to plastron is moderately broad, about 42% to 45% of carapace length. Head is small, blunt, with weakly bicuspid upper jaw. Shell is black above and below; skin of head, neck, and limbs is black to dark gray. White or cream mottling can occur on lower snout and lower jaw.

SIZE: Males attain a maximum length of nearly 90 cm SCL. Males range from 75 to ~118 cm CCL; females from 60 to 90 cm CCL.

OCCURRENCE: This species was divided into two populations on Isla San Cristóbal. The southern population is now extinct; the northern one persists with about 1,000 individuals.

HABITAT CHOICE: The habitat is more mesic, with scrubby woodlands edged by grasslands.

MISCELLANEA: *C. chathamensis*'s diet consists of grasses, forbs, cactus, and leaves of shrubs. Nesting begins in September, and the females deposit small clutches of four to six eggs. The small clutch size is characteristic of Galápagos tortoises, although eight to 10 eggs is the median number for most species.

Chelonoidis darwini   **Santiago Giant Tortoise**

Plate 34, Figure 29

APPEARANCE: Large to very large tortoise with shell intermediate between domed and saddleback. Rim of anterior shell opening is slightly or not at all upturned; posterior margin of shell is lightly flared and bluntly serrated. Bridge connecting carapace to plastron is moderately broad, about 40% to 43% of carapace length. Head is moderate-sized, blunt, with slightly hooked beak bordered by denticle on each side on upper jaw. Shell is gray to black above and gray below; skin of head, neck, and limbs is gray. Jaws are darker than the remainder of head and often have scattering of yellow spots.

SIZE: Males attain lengths of more than 140 cm CCL, usually ranging from 75 to 140 cm CCL; females from 55 to 95 cm CCL.

OCCURRENCE: This species lives on the volcano's slopes (200–700 m elevation) of Isla Santiago.

HABITAT CHOICE: The area occupied by this tortoise is predominately dry and rocky with scattered forest and scrub.

REPRODUCTION: The population nests between August and October. Clutch sizes range from four to 10 eggs. Precise incubation data are not available, although Galápagos tortoise incubation commonly ranges between 80 to 120 days.

MISCELLANEA: The current population possibly consists of 500 to 700 individuals. Their diet is largely made up of grasses, forbs, and cacti.

Chelonoidis ephippium   **Pinzón Giant Tortoise**

Plate 34

APPEARANCE: Large tortoise with saddleback shell. Rim of anterior shell opening is barely to strongly upturned. Posterior margin of shell has distinct outward and upward flare of marginal scutes and slightly serrated appearance. Bridge connecting carapace to plastron is moderately narrow, about 38% to 42% of carapace length. Head is small and blunt, with weakly hooked and bicuspid upper jaw. Shell is brownish gray above and below; skin of head, neck, and limbs is darker gray. Lower jaw and throat of male are commonly faded yellow.

SIZE: Among the Galápagos tortoises, *C. ephippium* is one of the smaller tortoises. The maximum reported size for a male is 87 cm SCL, although most males are 59 to 75 cm SCL with plastron lengths (PL) of 48 to 60 cm. Females are smaller, ranging in size from 46 to 68 cm SCL (maximum 77 cm SCL) and 38 to 55 cm PL.

OCCURRENCE: *C. ephippium* is a resident of Pinzón, where it occurs on the southern slope of that island's volcanic cone.

HABITAT CHOICE: The slopes of the volcano are covered with moderate dense thickets of shrubs and thorny trees interspersed with patches of grassland.

REPRODUCTION: These tortoises nest between August and December. Females lay clutches of two to eight eggs (median five) that require 85 to 120 days to hatch.

MISCELLANEA: The wild population has been restored by captive breeding and consists of approximately 500 subadult and adult tortoises. Presently, a naturally self-reproducing population seems unlikely owing to high egg and hatchling predation by the introduced Black Rats. This population is labeled *C. duncanensis* by some naturalists.

Chelonoidis guentheri   **Sierra Negra Giant Tortoise**
Plate 34

APPEARANCE: Large to very large tortoise with intermediate domed-saddleback shell. Rim of anterior shell opening is moderately low, its edge smoothly curved with no or slight upward tilt; posterior margin of shell is nearly vertical downward. Bridge connecting carapace to plastron is moderately broad. Head is moderate-sized and blunt, with weakly bicuspid upper jaw. Shell is grayish brown to black above and below; skin of head, neck, and limbs is gray. Lower jaw and throat of males are commonly faded yellow.

SIZE: Males attain a maximum length of about 120 cm SCL, although the usual range of male SCLs is 62 to 100 cm, 49 to 78 cm PL. Females are smaller, 60 to 74 cm SCL, 44 to 57 cm PL.

OCCURRENCE: *C. guentheri* lives on the lower slopes of the Volcan Sierra Negra, Isla Isabela.

HABITAT CHOICE: The dry slope of the volcano bears scrubby woodlands with more open areas of cactus and grasslands.

REPRODUCTION: Females lay modest clutches of eight to 17 eggs (median, nine) during the July-to-November nesting season.

MISCELLANEA: *C. guentheri* was driven to near extinct in the late 19th century by whalers. Fortunately, with the end of whale harvesting, the tortoise population slowly recovered during the 20th century and now numbers slightly over 2,000 individuals. These tortoises appear to survive mainly on a diet of grass and cactus pads.

Chelonoidis hoodensis   **Española Giant Tortoise**
Figure 29

APPEARANCE: Large tortoise with pronounced saddleback shell. Rim of anterior shell opening is slightly upturned; posterior margin of shell is nearly vertical downward with slight serrated appearance. Bridge connecting carapace to plastron is moderately narrow, about 37% to 38% of carapace length. Head is small and blunt, with bicuspid upper jaw. Shell is black above and below; skin of head, neck, and limbs is dark gray to black. Lower jaw and throat of adults are lightly mottled in white to faded yellow.

SIZE: *C. hoodensis* is the smallest Galápagos tortoise species. In the wild, males attain a maximum total length of approximately 75 cm SCL. Evidence for sexual dimorphism derives from growth in captivity where, in 20 years, a male attained 80 cm CCL (curved carapace length) and a female only 67 cm CCL.

OCCURRENCE: *E. hoodensis* is a resident of Isla Española, occurring inland on the slopes of the mountain.

HABITAT CHOICE: Isla Española is a dry, rocky island with a sparse vegetation of cactus, small thorny trees and shrubs, and patches of grass. It seldom receives rain and definitely is a desert island.

REPRODUCTION: Females nest from late June into December. Some individuals may deposit multiple clutches. Clutch size ranges from three to seven eggs. Incubation time varies from about 200 days for eggs laid in July to about 120 days for those laid later in November.

MISCELLANEA: The existence of *C. hoodensis* today is a conservation success story for the Charles Darwin Research Station. By the early 1960s, only three males and 12 females remained of this species. They were brought into captivity in 1965 to foster reproduction and to protect the resulting eggs and hatchlings. In 2000, 1,000 juveniles were released on Isla Española.

Chelonoidis microphyes   **Darwin's Giant Tortoise**

Not illustrated

APPEARANCE: Large to very large tortoise with domed shell. Rim of anterior shell opening is weakly upturned; posterior margin of shell is slightly downturned with smooth border. Bridge connecting carapace to plastron is moderately wide, about 46% to 48% of carapace length. Head is moderate-sized and blunt, with weakly hooked upper jaw bordered on each side by small denticle. Shell is brownish gray above and below; skin of head, neck, and limbs is gray.

SIZE: Adult males range from 74 to 103 cm SCL (~75–140 cm CCL) and 55 to 73 cm PL. Females are smaller, with an average SCL of about 64 cm (~55–95 cm CCL).

OCCURRENCE: This species lives in the southwestern lowlands and on the slopes of Volcan Darwin, Isla Isabela.

HABITAT CHOICE: The Darwin's Giant Tortoise lives in a dryer habitat than their relatives elsewhere on Isla Isabela, because Volcan Darwin is lower and lies in the rain shadow of Volcan Alcedo. Vegetation is mostly scrubby thickets of spiny trees and bushes with open spaces of grass bunches and cacti.

REPRODUCTION: In captivity, an endocrine study revealed a strong seasonal cycle in both males and females. Courtship and mating occurred in August to November and nesting from November through April, with a clutch size of eight to 17 eggs (with a mean of nine eggs).

MISCELLANEA: The present population size of *C. microphyes* is roughly 1,000 individuals.

Chelonoidis porteri   **La Caseta Giant Tortoise**

Plate 34

APPEARANCE: Large to very large tortoise with domed shell. Rim of anterior shell opening is smooth curve and not upturned; posterior margin of shell is flared, often slightly upturned, and serrated. Bridge connecting carapace to plastron is fairly broad, about 47% to 49% of carapace length. Head is small and blunt, with weakly hooked upper jaw bordered on each side by small denticle. Shell is black above and below; skin of head, neck, and limbs is dark gray to black.

SIZE: Males attain a maximum size of about 150 cm CCL with a range of ~85 to 150 cm CCL. Females are smaller than males, although they average larger than some other Galápagos species females, ranging from ~65 to 105 cm CCL.

Apalone spinifera   **Spiny Softshell**

Plate 35

APPEARANCE: Small to midsize freshwater turtle with nearly circular carapace. Surfaces of carapace and plastron are leathery, lacking horny scutes, and its edges are flexible. Light-brown to olive carapace typically has yellowish border and small light or dark spots scattered over entire surface; carapace's surface is lightly spiculate, particularly along anterior, slightly thickened rim. Plastron is light yellow to white and unmarked except for dark areas created by callosities of underlying plastral bones. Neck skin is smooth; yellow, black-edged stripe extends from in front of eye along side of head onto neck.

SIZE: Females average nearly two times larger than males, with males attaining sexual maturity early when they are as small as 8 to 10 cm SCL; females grow to ~18 cm SCL before beginning to mature. The range of adult body sizes depends on geography, generally with northern populations being smaller. Typically, females attain carapace lengths of 20 to 36 cm SCL (maximum 54 cm) and adult males 10 to 20 cm SCL. Hatchlings range from 30 to 44 mm SCL.

OCCURRENCE: In contrast to the two Asian softshell species, the spiny softshell is a recent immigrant (intentionally introduced in the 1990s) to O'ahu, Hawaii. The initial report derived from a small pond on a high school campus. It remains uncertain whether these softshells have established a breeding population, and they have not been reported from any other localities in the state. The native distribution of this turtle encompasses the entire Mississippi drainage system and westward into Texas, including the Rio Grande, and eastward to the Atlantic coast, although not into peninsular Florida.

HABITAT CHOICE: Spiny softshells are mainly river turtles within their natural North American distribution, although they occur largely in areas of slower moving water of small tributaries and adjacent oxbow lakes and ponds. It is this adaptation to slower waters that may enable spiny softshells to adapt to the small streams, ponds, and marshes of O'ahu.

REPRODUCTION: In the Hawaiin Islands, the nesting season is unknown. In North America, they generally nest from May through July. They dig their nests on pond and canal banks and deposit clutches of three to 39 (average 18) spherical, hard-shelled eggs. The eggs require from 10 to 12 weeks incubation before the hatchlings emerge.

MISCELLANEA: The Spiny Softshell is a catholic carnivore, eating the full variety of aquatic invertebrates and vertebrates.

Palea steindachneri   **Wattle-necked Softshell**
Plate 35

APPEARANCE: Midsize to large freshwater turtle with near circular carapace, broadest posteriorly. Surfaces of carapace and plastron are leathery and lack horny scutes, and have flexible edges. Brown to grayish-olive carapace bears thick ridge on its anterior margin, and ridge bears numerous blunt tubercles; remainder of carapace in adults is smooth. Plastron is light and unmarked, aside from dark areas external to underlying bony callosities of plastral bones. The name "wattle-necked" refers to shaggy cluster of large tubercles on top and sides of neck. Light-yellow spot lies at angle of jaw, bright in juveniles and fading with increasing age.

SIZE: Females average larger than males. Males attain a maximum length of ~36 cm SCL; adult females attain a maximum length of ~45 cm SCL. Hatchlings range from 25 to 34 mm SCL.

OCCURRENCE: This Asian softshell occurs naturally in southern China to northern Vietnam. Its presence in the waterways of Hawai'i, Kaua'i, Maui, and O'ahu probably began in the mid- to late 1880s with the importation of Asian foodstuffs for the Chinese community. Evidence indicates that multiple introductions occured over many years and that the geographic origins of these introductions were diverse.

HABITAT CHOICE: The Wattle-necked Softshell resides in slow-moving water, usually in areas of light to modest vegetation. In the Hawaiian Islands, it occurs predominantly in marshes and agricultural drainage canals.

REPRODUCTION: The Hawaiian Wattle-necks nest in June, and probably in other months as well. They dig their nests on pond and canal banks and deposit clutches of three to 28 spherical, hard-shelled eggs. The eggs require from seven to twelve weeks incubation before the hatchlings emerge.

MISCELLANEA: Wattle-necked Softshells are opportunistic carnivores and eat a large variety of invertebrate and vertebrate prey. They readily defend themselves by biting. With their long, flexible necks and sharp jaws, they must be handled carefully.

Pelodiscus sinensis **Chinese Softshell**

Plate 35

APPEARANCE: Midsize freshwater turtle with circular to broad, oblong carapace. Surfaces of carapace and plastron are leathery, lack horny scutes, and have flexible edges. Grayish-brown to grayish-olive carapace of adults is smooth and has slightly thickened anterior rim. Plastron is light colored and unmarked except for dark areas external to underlying bony callosities of plastral bones. Plastrons of hatchlings and juveniles have dark blotches. Neck skin is smooth and side of head and neck lack light-yellow postorbital stripe.

SIZE: *P. sinensis* is the smallest species of trionychids turtles. Females average larger than males. Adult females attain a maximum length of ~33 cm SCL; males a maximum of ~27 cm SCL. Hatchlings range from 24 to 30 mm SCL.

OCCURRENCE: As its common name indicates, this turtle derives from China. It has an extensive distribution from southeastern Siberia through eastern and southern China to central Vietnam. It is well established in the Hawaiian Islands of Kaua'i, Maui, and O'ahu, and was imported by the Chinese community in the mid- to late-1880s. The shipments of softshells probably derived from numerous and geographically distant Chinese populations. The Guam population came from Taiwan in 1977 as a commercial venture, which failed, and the turtles were released.

HABITAT CHOICE: The Chinese Softshell resides in slow-moving water, usually in areas of light to modest vegetation. In the Hawaiin Islands, it occurs predominantly in marshes and agricultural drainage canals.

REPRODUCTION: In the Hawaiian Islands, *P. sinensis* nests in late May through August. It digs its nests on pond and canal banks and deposits clutches of five to 48 spherical, hard-shelled eggs. The eggs require from seven to 12 weeks of incubation before the hatchlings emerge.

MISCELLANEA: A recent molecular study demonstrated that *P. sinensis* is not a single species, rather an evolutionary complex of at least five species. Scientific names are available for some of the species lineages but not for all. As yet, the genetic origin of the Hawaiian population has not been examined, and it is likely that the *P. sinensis* individuals released in Hawaii derived from more than a single Chinese population and perhaps also Japanese populations, and probably represent multiple species.

CROCODILIA: CROCODILIANS

Crocodilians are recognized instantly by everyone. Crocodilians, like sharks, fascinate because of the potential behavior of a few species to attack and eat humans, so everyone has a mental image of a "crocodile." The elongate, armored body and tail with robust, long-snouted head and fully toothed jaw are uniquely crocodilian among living vertebrates. There are three groups of these toothy jawed, armored reptiles: true crocodiles (Crocodylidae); gharials (Gavialidae); and alligators and caimans (Alligatoridae). All are semi-aquatic, spending much of their lives in water although emerging regularly to bask along the water's edge and all lay eggs on shore, either in detritus nests or in holes dug by the females. Crocodilians occur throughout the tropical to semitropical waterways of the world, living in variety of habitats from small jungle streams to marshes and swamps, and from ponds and lakes to the largest tropical rivers. Each species is adapted to a specific set of aquatic habitats and the prey therein. All are stealth predators. They capture their prey in the water or at the water's edge, and they consume their prey in the water.

Crocodylidae: Crocodiles

Crocodiles occur pantropically. They live in the waterways of all continents. Most crocodiles are freshwater inhabitants, but two—the American Crocodile and the Saltwater Crocodile—are predominantly coastal marine in estuaries and mangrove forest. All crocodiles share a unique tooth feature that allows them to be identified from a distance. When the jaw is closed, the fourth mandibular (lower jaw) tooth lies outside the mouth and is the only tooth visible. No teeth are outside the closed mouth of caiman and alligators, and many teeth bristle outside the closed mouth of gharials. A single species lives among the western Pacific islands, the Saltwater Crocodile.

Crocodylus porosus   **Saltwater Crocodile**

Plate 35, Figure 30

APPEARANCE: Large to very large crocodilian. Its color ranges from golden tan to dark gray, and most individuals have dark lateral spots or bars. Color is brightest in hatchlings and juveniles and becomes muted

FIGURE 30. The arrangement of dorsal armor in crocodiles is diagnostic for most species. *Crocodylus porosus* is the only species to have an isolated cluster of four to six large cervical osteoderms amid a field of tiny tuberculate scales. Rarely one or two small osteoderms occur in the nuchal area. The dorsal trunk shield of osteoderms begins over the shoulders.

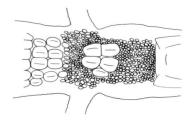

with age. Dorsal armor extends continuously from above shoulders onto tail. Horny scales are underlain by bony plates. Scales are rounded with central keel, regularly arranged in transverse rows, and typically not abutting one another on their sides. Back of head to shoulders has single shield of four large, keeled scales arranged in transverse pairs and usually bordered on each side by one (occasionally two) smaller, keeled scale.

SIZE: Males average larger than females. The maximum confirmed total length is 6.2 m for a New Guinean *C. porosus*; lengths of seven meters have been reported for individuals from India and Sri Lanka but those lengths are unconfirmed. A recent survey of the Palauan population yielded the following size data: females from 38 to 64 cm SVL, 112 to 264 cm total length; males from 40 to 82 cm SVL, 110 to 230 cm total length. The surveyors saw, but did not capture, individuals to about 3.3 m. Hatchlings average about 29 cm in total length.

OCCURRENCE: "Salties," as *C. porosus* is colloquially named, have an enormous distribution from southwest coastal India and Sri Lanka eastward through Asia and the Philippines to Palau and southward into tropical Australia. Within the Pacific limits of this book, *C. porosus* has long-established populations in the Santa Cruz Islands, Vanuatu, and Palau. Vagrant individuals have been reported in Fiji and throughout Federated States of Micronesia from Yap to Pohnpei. The most remote waif was a juvenile that reached Nauru in 1994.

HABITAT CHOICE: *C. porosus* can and does tolerate saltwater and makes long-distance dispersal movement across the open sea; nevertheless, it is principally a coastal animal of brackish and freshwater. Along the coasts, it prefers areas of mangroves and small tidal streams with overhanging vegetation. It also resides in freshwater rivers and streams; in some instances, it lives kilometers upstream from saltwater.

REPRODUCTION: Saltwater crocodiles are mound-nesters if sufficient vegetable detritus is available. Like all crocodilians, females practice parental care with behavior such as guarding the mound-nest, transporting the hatchlings from the nest to water, staying with or nearby the hatchlings, and actively protecting them from predators for days to weeks. Fifty to 60 elliptical, flexible-shelled eggs are deposited in a hole dug in the middle of the mound. The eggs incubate for 80 to 90 days. When the young are ready to hatch, they begin to chirp. The female uncovers the eggs and assists the hatchlings to break through and exit the eggshells. She carries each hatchling, individually, to the water.

MISCELLANEA: *C. porosus* is a carnivore from hatching to death. It captures invertebrate and vertebrate prey appropriate to its size,—that is, any animal that it can catch and subdue. This catholic carnivory makes it dangerous to humans. Although it is wary of humans and stays hidden when people are near their residences, an unwary human is simply another food item and will be captured if the opportunity is offered.

In spite of the potential negative interaction with humans, population size can be substantial. The previously mentioned survey and assessment (2003) of the Palauan crocodiles indicated that the resident population contained 500 to 750 juvenile and adult crocodiles.

Acknowledgments

My research on the Pacific herpetofauna began in 1971. Throughout the decades since then, many people have assisted and encouraged me. I offer a heartfelt thank you to all. More of you are unnamed than named herein, and most are unlikely to see or use this guide. That does not lessen my appreciation and gratuity for your assistance, whether it was a moment's assistance with directions or several days accompanying me in my search for lizards. I have experienced a wealth of friendliness and helpfulness. Thank you all!

With the readers' forbearance, I identify the role of a few supporters of my Pacific research in the form of an autobiographic sketch. My—really our (my wife Pat's and mine)—Pacific adventures began in 1970 with an inquiry from Max Downes (director of the PNG Wildlife Laboratory) during his initial visit to Washington, DC. Discovering the wealth of crocodilian literature in the Smithsonian and the Library of Congress, he casually asked Jim Peters (curator in charge of our reptiles division) and me whether we might know someone who was willing to do a house swap. I expressed a definite interest. That evening, I ask Pat whether she shared my interest. Without hesitation, she said "Why not?" The next morning, Jim Peters wholeheartedly agreed with our decision, even though my six-month absence from the division would leave him shorthanded.

On September 30, 1971, Pat and I arrived at Jackson Airport, Port Moresby, Papua New Guinea (PNG), with son Jon (10 years old) and

Erin (2 years). Eric Lindgren, acting director of the Moitaka Wildlife Laboratory, met us and took us to the Downes' house in Boroko, PNG. Eric and his wife, Del, were incredibly helpful in introducing this naive family to life in the tropics. We quickly adjusted to the Moresby lifestyle in our Boroko home and at the Wildlife Lab in Moitaka. My research focus was anuran locomotion, which was slow to get started because we arrived during the dry season. The first heavy rainstorm was a month away, so I began to explore locally for lizards. John Pippet, Wildlife Lab station manager and an enthusiast of long-horned beetles, joined me on many of my ventures into the savannah and nearby rainforest. He and Eric suggested a Cane Toad census and sampling project, which became part of our monthly routine. The local forays introduced me to two lizards (*Nactus* and *Carlia*) that have been part of my Pacific research ever since. Our five months in PNG introduced me to the diversity of tropical landscapes and herpetofaunas. I caught the Pacific research virus, which has cycled in and out of my professional career since that time.

Were it not for Vic Springer, I might have gotten over the virus. It was at low ebb after a decade spent in exhibit production and the chairmanship of the museum's vertebrate zoology department. In 1981, Vic invited me to join him and our fish division colleagues in an expedition to Tonga in the spring of 1982. No one is more attentive to the details and logistics of fieldwork planning than Vic. Because of his attentiveness, our field equipment was whisked off the dock at Nuku'alofa a day or two before Tongatapu and nearby islands were flatten by a huge cyclone. Our gear was routed to Suva, Fiji. Dr. A. D. Lewis of the Lami Fisheries Laboratory of the Fijian Ministry of Primary Industry greatly assisted Vic in arranging nearshore fish surveys among the Fijian islands, most notably in the central and southern Lau group. I was along for the boat ride and was regularly dropped ashore while they dove for fishes. I also met John Gibbons, who became a friend during a couple of herping excursions. He and his colleagues at the Department of Biology (University of the South Pacific) were always willing hosts, both then and on subsequent visits to Fiji through the 1990s. I joined Vic in 1986 for an expedition to Rotuma, again arranged through the Fisheries Laboratory and Dr. Lewis. Following that Rotuma expedition, Lily and John Gibbons and their children provided a home away from home; a delightful stay that became a bittersweet memory with their drowning deaths in a boating accident later that year. A special thank you is owed to Dr. Niumaia Tabunakawai of Primary Industries and subsequent government departments for his encouragement of our herpetological fieldwork in Fiji and his assistance with the export of voucher specimens during my numerous visits.

After a hiatus of several years, my Pacific fieldwork began anew in 1992. Ivan Ineich, curator at the Paris natural history museum, joined me in September and October in Fiji, Tonga, and Samoa. In 1993, Robert Fisher, then a graduate student at UC-Davis, met me in Efate, Vanuatu, for a visit that was foreshortened owing to my choice of a government official, whose mishandling of my documents caused a rejection of my research request by the minister of wildlife. The abbreviated Vanuatu visit led to a longer field season in Fiji, where we were joined by my wife Pat—her first return to the Pacific since our time in PNG. Jeff Williams organized a nearshore survey of fishes for Tonga in October 1993, and invited me to join him and other fish division colleagues for a month-long trip. Owing to the size of the survey boat, we were land based with daily dives from the boat by the ichthyologists. This trip provided me with the opportunity to visit the Va'vau and Ha'apai island groups, in addition to Tongatapu. When the ichthyologists returned stateside, I awaited Pat's arrival in Nadi. When she arrived, we immediately departed for fieldwork in Vanuatu. In subsequent years, Pat and I worked in Rapa Nui and Rarotonga, Cook Islands. Again through the courtesy of Jeff Williams, I was able to visit islands of the Santa Cruz group, Solomon Islands. Our 1998 "adventures" on the interisland freighter, the *Butai*, are strong memories.

In addition to fieldwork, my Pacific research has drawn heavily on the collections of many museums. I wholeheartedly thank the curators and collections management staff of the following museums for lending specimens or hosting me, or both: the American Museum of Natural History, the Auckland Institute and Museum, Australian Museum – Sydney, the Bernice P. Bishop Museum, the Californian Academy of Sciences, the Carnegie Museum of Natural History, the Field Museum of Natural History, the Florida Museum of Natural History, the Institut Royal des Sciences Naturelle de Belgique, the Los Angeles County Museum, the Louisiana State Museum of Natural Sciences, the Museum für Naturkunde – Berlin, the Museum of Comparative Zoology, the Muséum national d'Historie naturelle, the Natural History Museum – London, the Naturalis-Nationaal Natuurhistorisch Museum, the Naturhistorisches Museum – Wien, the Queensland Museum of Natural History, the San Diego Museum of Natural History, the Senckenberg Forschunginstitut und Naturmuseum, the South

Australian Museum, the Western Australian Museum, the Zoologisches Forschungsinstitut und Museum Alexander Koenig, and the Zoologisches Sammlung des Bayerischen Staates. I offer special thanks to Carla Kishinami, Colin McCarthy, José Rosado, Ross Sadlier, and Jens Vindum, who have been subjected to numerous visits and many special requests over many years; they have steadfastly provided outstanding hospitality and service. While my present and past colleagues in the Division of Amphibians and Reptiles remain unnamed here, I thank them for their excellent assistance and collegiality. I also offer a special thank you to my colleagues in the natural history branch of the Smithsonian Institution Library (SIL). Our library resources are outstanding and the assistance that I have received from the SIL staff is similarly outstanding.

Travel for fieldwork and the study of museum collection requires financial support. I sincerely thank the Max & Victoria Dreyfus Foundation, the National Geographic Society, and the National Science Foundation, which have assisted my Pacific research. My biggest supporter has been the National Museum of Natural History, Smithsonian Institution. Aside from being my employer for 38 years, the Museum continues to provide me with research resources and space. It has also supported research travel through a variety of programs, such as the Fluid Research Fund, Research Opportunity Funds, the Smithsonian Scholarly Research Fund, and others. I am most appreciative of all resources that the Smithsonian has and continues to supply.

ASSISTANCE WITH THIS GUIDE

Although errors probably persist, they are far fewer owing to the critical reading by friends and colleagues. I must single out for a special thank you Steve Busack and Hope Steele, who reviewed the text in its entirety, corrected mistakes, and adjusted my quirky phrasing and word selection. Other reviewers include Pat Zug, Robert Fisher, and Ron Crombie, who commented on early drafts of partial sections; Rachel Hensen and Eli Wostl, who provided comments on the frog species accounts; Bayard Brattstrom, who provided comments on the Revillagigedo squamate accounts; Don Buden, on the Micronesia regional checklist and lizard species accounts; Hal Cogger, on the seasnakes and seakrait accounts; Carl Ernst, on the entire turtle section; Jack Frazier, on select seaturtle accounts; Ali Hamilton, on the Southwest Pacific gecko and skink accounts; Ivan Ineich, on the South

Central Pacific and Eastern Pacific regional checklists as well as assorted lizard and seasnake accounts; Fred Kraus, on the Hawaiian frog and lizard accounts; Kenney Krysko, on the alien species accounts; Omar Torres-Carvajal, on the *Phyllodactylus* and *Microlophus* accounts; and Addison Wynn on the *Ramphotyphlops* species accounts.

Robert Thomas shared his extensive Galápagos colubrid snakes data with me. Carl Ernst and John Iverson similarly provided me with all their morphometric data on *Mauremys reevesii*.

It has been my good fortune to receive color images of Pacific herps and permission to use them in this guide from many talented natural- ists. I thank the following individuals for their images: Chris Austin, Bayard Brattstrom, Don Buden, Hal Cogger, Ron Crombie, Indraneil Das, Kevin de Queiroz, Kevin Enge, Carl Ernst, Margie Falanruw, Robert Fisher, Tom Fritts, Steve Gotte, Ali Hamilton, Blair Hedges, Ivan Ineich, Fred Kraus, Kenney Krysko, Björn Lardner, Thomas Leuteritz, Mateo Lopez Victoria, Mike McCoy, Roy McDiarmid, Dan Moen, Hans-Peter Phillip, Greg Pregill, Bob Reynolds, Santiago Ron, Paddy Ryan, Heidi Snell, John Tashijian, Omar Torres-Carvajal, Robert G. Tuck, Jens Vindum, Tony [Anthony H.] Whitaker, Elijah Wostl, and Bob Zappalorti. Their specific contributions are identified in the Illustration Sources appendix. Ulrich Manthey supplied images of the head of the type of *Hypsilurus godeffroyi* so that we could include a line drawing of this species.

Last but not least, I thank the illustrators who have assisted me in my Pacific research by providing lining drawings of numerous Pacific species. Molly Dwyer Griffin and Ted Kahn provided images specifi- cally for this guide. Finally, a special thank you goes to my wife Pat for her support from the beginning of this guide to the final preparation of its indexes.

Checklist of Pacific Island Amphibians and Reptiles

Bufonidae: Toads

☐ *Rhinella marina*   Cane Toad

Ceratobatrachidae: Pacific Forest Frogs

☐ *Platymantis pelewensis*   Palau Frog

☐ *Platymantis vitiana*   Fiji Ground Frog

☐ *Platymantis vitiensis*   Fiji Treefrog

Dendrobatidae: Poison Frogs

☐ *Dendrobates auratus*   Green and Black Poison-dart Frog

Dicroglossidae: Fork-tongued Frogs

☐ *Fejervarya cancrivora*   Crab-eating Frog

☐ *Fejervarya "limnocharis"*   Paddy Frog

Eleutherodactylidae: Rain Frogs

☐ *Eleutherodactylus coqui*   Coquí

☐ *Eleutherodactylus planirostris*   Greenhouse Frog

Hylidae: Treefrogs

☐ *Litoria aurea* Green and Gold Bellfrog

☐ *Litoria fallax* Eastern Dwarf Treefrog

☐ *Scinax quinquefasciatus* Five-lined Snouted Treefrog

Microhylidae: Narrow-mouthed Toads

☐ *Microhyla pulchra* Marbled Pygmy Frog

Ranidae: Water Frogs

☐ *Glandirana rugosa* Japanese Wrinkled Frog

☐ *Hylarana guentheri* Brown and Tan Amoy Frog

☐ *Lithobates catesbeianus* American Bullfrog

Rhacophoridae: Afro-Asian Treefrogs

☐ *Polypedates braueri* Taiwan Whipping Frog

REPTILIA: SQUAMATA (LIZARDS)

Gekkota

Gekkonidae: Geckos

☐ *Gehyra brevipalmata* Palau Gecko

☐ *Gehyra georgpotthasti* Vanautu Giant Gecko

☐ *Gehyra insulensis* Pacific Stump-toed Gecko

☐ *Gehyra oceanica* Oceania Gecko

☐ *Gehyra vorax* Fiji Giant Gecko

☐ *Gekko gecko* Tokay

☐ *Gekko vittatus* Melanesia Ghost Gecko

☐ *Gekko* nsp Palau Ghost Gecko

☐ *Hemidactylus frenatus* Common House Gecko

☐ *Hemidactylus garnotii* Fox Gecko

☐ *Hemiphyllodactylus ganoklonis* Palau Slender Gecko

☐ *Hemiphyllodactylus typus* Indo-Pacific Slender Gecko

☐ *Lepidodactylus buleli*   Vanuatu Ant-nest Gecko

☐ *Lepidodactylus euaensis*   'Eua Forest Gecko

☐ *Lepidodactylus gardineri*   Rotuma Forest Gecko

☐ *Lepidodactylus guppyi*   Solomon Forest Gecko

☐ *Lepidodactylus lugubris*   Mourning Gecko

☐ *Lepidodactylus manni*   Fiji Forest Gecko

☐ *Lepidodactylus moestus*   Micronesia Flat-tailed Gecko

☐ *Lepidodactylus oligoporus*   Mortlock Forest Gecko

☐ *Lepidodactylus paurolepis*   Palau Barred Gecko

☐ *Lepidodactylus tepukapili*   Tuvalu Forest Gecko

☐ *Lepidodactylus vanuatuensis*   Vanuatu Forest Gecko

☐ *Lepidodactylus* nsp   Central Pacific Beach Gecko

☐ *Nactus multicarinatus*   Melanesia Slender-toed Gecko

☐ *Nactus pelagicus* `   Pacific Slender-toed Gecko

☐ *Perochirus ateles*   Micronesia Saw-tailed Gecko

☐ *Perochirus guentheri*   Vanuatu Saw-tailed Gecko

☐ *Perochirus scutellatus*   Giant Saw-tailed Gecko

☐ *Phelsuma grandis*   Madagascar Giant Daygecko

☐ *Phelsuma guimbeaui*   Orange-spotted Daygecko

☐ *Phelsuma laticauda*   Golddust Daygecko

Phyllodactylidae: Leaf-toed Geckos

☐ *Phyllodactylus barringtonensis*   Santa Fé Leaf-toed Gecko

☐ *Phyllodactylus baurii*   Pinta Leaf-toed Gecko

☐ *Phyllodactylus darwini*   Darwin's Leaf-toed Gecko

☐ *Phyllodactylus galapagensis*   Galápagos Leaf-toed Gecko

☐ *Phyllodactylus gilberti*   Wolf Leaf-toed Gecko

☐ *Phyllodactylus leei*   San Cristóbal Leaf-toed Gecko

☐ *Phyllodactylus reissii*   Guayaquil Leaf-toed Gecko

☐ *Phyllodactylus transversalis*   Malpelo Leaf-toed Gecko

Sphaerodactylidae: Miniature Geckos

☐ *Gonatodes caudiscutatus*   Shield-headed Gecko

☐ *Sphaerodactylus pacificus*   Cocos Pygmy Gecko

Iguania

Dactyloidae: Anoles

☐ *Anolis agassizii*   Malpelo Anole

☐ *Anolis carolinensis*   Green Anole

☐ *Anolis equestris*   Knight Anole

☐ *Anolis sagrei*   Brown Anole

☐ *Anolis townsendi*   Cocos Anole

Iguanidae: Iguanas

☐ *Amblyrhynchus cristatus*   Marine Iguana

☐ *Brachylophus bulabula*   Fiji Banded Iguana

☐ *Brachylophus fasciatus*   Lau Banded Iguana

☐ *Brachylophus vitiensis*   Fiji Crested Iguana

☐ *Conolophus marthae*   Pink Land Iguana

☐ *Conolophus pallidus*   Santa Fé Land Iguana

☐ *Conolophus subcristatus*   Galápagos Land Iguana

☐ *Iguana iguana*   Green Iguana

Phrynosomatidae: Spiny Lizards and Relatives

☐ *Urosaurus auriculatus*   Socorro Treelizard

☐ *Urosaurus clarionensis*   Clarion Treelizard

Tropiduridae: Lava Lizards

☐ *Microlophus albemarlensis*   Galápagos Lava Lizard

☐ *Microlophus bivittatus*   San Cristóbal Lava Lizard

☐ *Microlophus delanonis*   Española Lava Lizard

☐ *Microlophus duncanensis*   Pinzón Lava Lizard

☐ *Microlophus grayii*   Floreana Lava Lizard

☐ *Microlophus habelii*   Marchena Lava Lizard

☐ *Microlophus pacificus*   Pinta Lava Lizard

Scincomorpha

Eugongylidae: Skinks

☐ *Carlia ailanpalai*   Admiralty Brown Skink

☐ *Carlia tutela*   Halmahera Brown Skink

☐ *Cryptoblepharus eximius*   Fiji Snake-eyed Skink

☐ *Cryptoblepharus nigropunctatus*   Bonin Snake-eyed Skink

☐ *Cryptoblepharus novohebridicus*   Vanuatu Snake-eyed Skink

☐ *Cryptoblepharus poecilopleurus*   Oceania Snake-eyed Skink

☐ *Cryptoblepharus rutilus*   Palau Snake-eyed Skink

☐ *Emoia adspersa*   Striped Small-scaled Skink

☐ *Emoia aneityumensis*   Anatom Treeskink

☐ *Emoia arnoensis*   Micronesia Black Skink

☐ *Emoia atrocostata*   Seaside Skink

☐ *Emoia boettgeri*   Micronesia Spotted Skink

☐ *Emoia caeruleocauda*   Pacific Blue-tailed Skink

☐ *Emoia campbelli*   Vitilevu Mountain Treeskink

☐ *Emoia concolor*   Fiji Slender Treeskink

☐ *Emoia cyanogaster*   Green-bellied Vineskink

☐ *Emoia cyanura*   White-bellied Copper-striped Skink

☐ *Emoia erronan*   Erronan Treeskink

☐ *Emoia impar*   Dark-bellied Copper-striped Skink

☐ *Emoia jakati*   Papua Five-striped Skink

☐ *Emoia lawesii*   Olive Small-scaled Skink

☐ *Emoia mokolahi*   Tonga Robust Treeskink

☐ *Emoia mokosariniveikau*   Vanualevu Slender Treeskink

☐ *Emoia nigra*   South Pacific Black Skink

☐ *Emoia nigromarginata*   Vanuatu Coppery Vineskink

☐ *Emoia oriva*   Rotuma Barred Treeskink

☐ *Emoia parkeri*   Fiji Copper-headed Skink

☐ *Emoia ponapea*   Pohnpei Skink

☐ *Emoia rufilabialis*   Red-lipped Striped Skink

☐ *Emoia samoensis*   Pacific Robust Treeskink

☐ *Emoia sanfordi*   Toupeed Treeskink

☐ *Emoia slevini*   Mariana Skink

☐ *Emoia taumakoensis*   Taumako Skink

☐ *Emoia tongana*   Polynesia Slender Treeskink

☐ *Emoia trossula*   Fiji Barred Treeskink

☐ *Emoia tuitarere*   Rarotonga Treeskink

☐ *Eugongylus albofasciolatus*   Barred Recluse Skink

☐ *Eugongylus* nsp   Palau Recluse Skink

☐ *Tachygyia microlepis*   Small-scaled Giant Skink

Lygosomidae: Skinks

☐ *Caledoniscincus atropunctatus*   Speckled Litter Skink

☐ *Lamprolepis smaragdina*   Emerald Treeskink

☐ *Lampropholis delicata*   Garden Skink

☐ *Leiolopisma alazon*   Ono-i-Lau Ground Skink

Mabuyidae: Multiple-keeled Skinks

☐ *Eutropis multicarinata*   Micronesia Multi-keeled Sunskink

Sphenomorphidae: Twilight Skinks

☐ *Lipinia leptosoma*   Pandanus Moth Skink

☐ *Lipinia noctua*   Pacific Moth Skink

☐ *Prasinohaema virens*   Green-blooded Vineskink

☐ *Sphenomorphus scutatus*   Palau Ground Skink

☐ *Sphenomorphus solomonis*   Solomon Ground Skink

Anguimorpha

Diploglossidae: Galliwasps and Relatives

☐ *Diploglossus millepunctatus*   Malpelo Galliwasp

Varanidae: Monitors and Goannas

☐ *Varanus "indicus"*   Pacific Monitor

Other Lizards, most Aliens

Agamidae: Dragons and Relatives

☐ *Hypsilurus godeffroyi*   Papau Angle-headed Lizard

Chamaeleonidae:Chameleons

☐ *Chamaeleo calyptratus*   Veiled Chameleon

☐ *Trioceros jacksonii*   Jackson's Chameleon

REPTILIA: SQUAMATA (SNAKES)

Typhlopidae: Blindsnakes

☐ *Ramphotyphlops acuticaudus*   Palau Blindsnake

☐ *Ramphotyphlops adocetus*   Ant Atoll Blindsnake

☐ *Ramphotyphlops braminus*   Brahminy Blindsnake

☐ *Ramphotyphlops depressus*   Melanesia Blindsnake

☐ *Ramphotyphlops hatmaliyeb*   Ulithi Blindsnake

☐ *Ramphotyphlops* nsp   Taveuni Blindsnake

Boidae: Boas

☐ *Candoia bibroni*   Pacific Treeboa

☐ *Candoia paulsoni*   Melanesia Bevel-nosed Boa

☐ *Candoia superciliosa*   Palau Bevel-nosed Boa

Colubridae: Common Snakes

☐ *Alsophis biseralis*   Galápagos Racer

☐ *Antillophis slevini*   Galápagos Banded Snake

☐ *Antillophis steindachneri*   Galápagos Striped Snake

☐ *Boiga irregularis*   Brown Treesnake

☐ *Dendrelaphis salomonis*   Solomon Treesnake

☐ *Dendrelaphis striolatus*   Palau Treesnake

☐ *Masticophis anthonyi*   Clarion Racer

☐ *Philodryas hoodensis*   Española Racer

Elapidae: Cobras and Allies

☐ *Hydrophis coggeri*   Pacific Yellow-banded Seasnake

☐ *Hydrophis pacificus*   Pacific Faint-banded Seasnake

☐ *Laticauda colubrina*   Yellow-lipped Seakrait

☐ *Laticauda frontalis*   Vanuatu Yellow-lipped Seakrait

☐ *Laticauda laticaudata*   Dark-lipped Seakrait

☐ *Laticauda schistorhyncha*   Central Pacific Seakrait

☐ *Ogmodon vitianus*   Fiji Bola

☐ *Pelamis platura*   Yellow-bellied Seasnake

Homalopsidae: Mudsnakes

☐ *Cerberus dunsoni*   Palau Dog-faced Mudsnake

REPTILIA: TESTUDINES

Chelidae: Sideneck Turtles

☐ *Chelodina longicollis*   Eastern Long-necked Turtle

Cheloniidae: Hard-shelled Seaturtles

☐ *Caretta caretta*   Loggerhead Seaturtle

☐ *Chelonia mydas*   Green Seaturtle

☐ *Eretmochelys imbricata*   Hawksbill Seaturtle

☐ *Lepidochelys olivacea*   Olive Ridley Seaturtle

Dermochelyidae: Leatherback Seaturtles

☐ *Dermochelys coriacea*   Leatherback Seaturtle

Chelydridae: Snapping Turtles

☐ *Chelydra serpentina*   Snapping Turtle

Emydidae: Pondturtles

☐ *Trachemys scripta*   Red-eared Slider

Geoemydidae: Pondturtles

☐ *Mauremys reevesii*   Chinese Three-keeled Pondturtle

☐ *Mauremys sinensis*   Chinese Stripe-necked Pondturtle

Testudinidae: Tortoises

☐ *Chelonoidis abingdonii*   Pinta Giant Tortoise

☐ *Chelonoidis becki*   Wolf Giant Tortoise

☐ *Chelonoidis chathamensis*   San Cristóbal Giant Tortoise

☐ *Chelonoidis darwini*   Santiago Giant Tortoise

☐ *Chelonoidis ephippium*   Pinzón Giant Tortoise

☐ *Chelonoidis guentheri*   Sierra Negra Giant Tortoise

☐ *Chelonoidis hoodensis*   Española Giant Tortoise

☐ *Chelonoidis microphyes*   Darwin's Giant Tortoise

☐ *Chelonoidis porteri*   La Caseta Giant Tortoise

☐ *Chelonoidis vandenburghi*   Alcedo Giant Tortoise

☐ *Chelonoidis vicina*   Cerro Azul Giant Tortoise

☐ *Chelonoidis sp*   Cerro Fatal Giant Tortoise

Trionychidae: Softshell Turtles

☐ *Apalone spinifera*   Spiny Softshell

☐ *Palea steindachneri*   Wattle-necked Softshell

☐ *Pelodiscus sinensis*   Chinese Softshell

REPTILIA: CROCODILIA

Crocodylidae: Crocodiles

☐ *Crocodylus porosus*   Saltwater Crocodile

Appendix: Sources for Illustrations

Amphibia: Anura

Bufonidae

Rhinella marina	Plate 4A: G.R. Zug [Hawaiian Islands, Kaua'i] female
Rhinella marina	Plate 4B: Elijah Wostl [Guam] male

Ceratobatrachidae

Platymantis pelewensis	Plate 1A: Elijah Wostl [Palau]
Platymantis pelewensis	Plate 1B: Christopher C. Austin [Palau]
Platymantis vitiana	Plate 1E: G.R. Zug [Fiji, Ovalau]
Platymantis vitiana	Plate 1F: G.R. Zug [Fiji, Ovalau]
Platymantis vitiensis	Plate 1C: G.R. Zug [Fiji, Viti Levu]
Platymantis vitiensis	Plate 1D: Paddy Ryan[Fiji, Viti Levu]

Dendrobatidae
Dendrobates auratus — Plate 4C: Fred Kraus [Hawaii, O'ahu]

Dicroglossidae
Fejervarya cancrivora — Plate 3A: Elijah Wostl [Guam]
Fejervarya "limnocharis" — Plate 3B: G.R. Zug [Guam]

Eleutherodactylidae
Eleutherodactylus coqui — Plate 4D: S. Blair Hedges [Puerto Rico]
Eleutherodactylus planirostris — Plate 4E: Kenneth L. Krysko [Florida]

Hylidae
Litoria aurea — Plate 2A: G.R. Zug [Vanuatu, Efate]
Litoria aurea — Plate 2B: A.H. Whitaker [New Caledonia]
Litoria fallax — Plate 2C: Elijah Wostl [Guam]
Litoria fallax — Plate 2D: Elijah Wostl [Guam]
Scinax quinquefasciatus — Plate 2E: Santiago Ron [Ecuador, Río Palenque]

Microhylidae
Microhyla pulchra — Plate 4F: G.R. Zug [Guam]

Ranidae
Glandirana rugosa — Plate 3D: G.R. Zug [Hawaii, Kaua'i]
Hylarana guentheri — Plate 3C: G.R. Zug [Guam]
Lithobates catesbeianus — Plate 3E: R.G. Tuck [Maryland, C&O Canal]

Rhacophoridae
Polypedates braueri — Plate 2F: G.R. Zug [Guam]

Reptilia: Squamata (Lizards)
Gekkonidae
Gehyra brevipalmata — Plate 5A: Gregory Pregill [Palau]
Gehyra insulensis — Plate 5B: Robert N. Fisher [Fiji?]

Gehyra oceanica	Plate 5C: G.R. Zug [Fiji, Taveuni]
Gehyra oceanica	Plate 5D: A.H. Whitaker [unknown]
Gehyra vorax	Plate 5E: Paddy Ryan [Fiji]
Gekko gecko	Plate 6C: Jens Vindum [Myanmar]
Gekko vittatus	Plate 6A: Mike McCoy [Solomon Islands]
Gekko nsp-Palau	Plate 6B: Gregory Pregill [Palau]
Hemidactylus frenatus	Plate 7A: G.R. Zug [Fiji, Viti Levu]
Hemidactylus garnotii	Plate 7B: G.R. Zug [Cook Islands, Rarotonga]
Hemiphyllodactylus ganoklonis	Plate 7C: Ronald I. Crombie [Palau]
Hemiphyllodactylus typus	Plate 7D: A.H. Whitaker [New Caledonia, Rivière Tontouta]
Lepidodactylus buleli	Plate 8D. Ivan Ineich [Vanuatu, Espiritu Santo]
Lepidodactylus euaensis	Plate 8A: G.R. Zug [Tonga, ʻEua]
Lepidodactylus gardineri	Plate 8B: Paddy Ryan [Rotuma]
Lepidodactylus guppyi	Plate 8E: G.R. Zug [Solomon Islands, Duff Island]
Lepidodactylus lugubris, clone A	Plate 9A: G.R. Zug [Tonga, Tongatabu]
Lepidodactylus lugubris, clone B	Plate 9B: G.R. Zug [Hawaii, Kauaʻi]
Lepidodactylus lugubris, clone C	Plate 9C: G.R. Zug [Tonga, Tongatabu]
Lepidodactylus lugubris, clone E	Plate 9D: G.R. Zug [Tonga, ʻEua]
Lepidodactylus manni	Plate 8C: G.R. Zug [Fiji, Viti Levu]
Lepidodactylus moestus	Plate 9E: Elijah Wostl [Palau, Koror]
Lepidodactylus vanuatuensis	Plate 8F: G.R. Zug [Vanuatu, Espiritu Santo]
Nactus multicarinatus	Plate 10A: G.R. Zug [Vanuatu, Espiritu Santo]
Nactus pelagicus	Plate 10B: G.R. Zug [Guam]
Perochirus ateles	Plate 6D: Christopher C. Austin [Federated States of Micronesia, Chuuk]
Phelsuma grandis	Plate 10E: Kenneth L. Krysko [Florida: Little Torch Key]
Phelsuma guimbeaui	Plate 10C: John H. Tashjian [captive]
Phelsuma laticauda	Plate 10D: Gregory Pregill [Hawaii]

Phyllodactylidae

Phyllodactylus barringtonensis	Plate 11B: Omar Torres Carvajal [Galápagos, Santa Fé]
Phyllodactylus baurii	Plate 11C: Omar Torres Carvajal [Galápagos]
Phyllodactylus galapagensis	Plate 11D: Omar Torres Carvajal [Galápagos]
Phyllodactylus reissii	Plate 11E: Omar Torres Carvajal [Galápagos]
Phyllodactylus transversalis	Plate 11F: Mateo Lopez-Victoria [Malpelo]

Sphaerodactylidae

Gonatodes caudiscutatus	Plate 11A: Roy W. McDiarmid [Ecuador, mainland]

Dactyloidae

Anolis agassizii	Plate 15A: Mateo Lopez Victoria [Malpelo] male
Anolis agassizii	Plate 15B: Mateo Lopez Victoria [Malpelo] female
Anolis carolinensis	Plate 15C: G.R. Zug [Guam]
Anolis equestris	Plate 15D: S. Blair Hedges [Cuba]
Anolis sagrei	Plate 15E: Fred Kraus [Hawaii]
Anolis sagrei	Plate 15F: Carla Kishinami [Hawaii]

Iguanidae

Amblyrhynchus cristatus	Plate 13A: Kevin de Queiroz [Galápagos, San Cristóbal]
Amblyrhynchus cristatus	Plate 13B: Kevin de Queiroz [Galápagos, Santa Cruz]
Brachylophus bulabula	Plate 12A: Robert N. Fisher [Fiji, Vanua Levu]
Brachylophus fasciatus	Plate 12B: Robert N. Fisher [Fiji, Lau Islands]
Brachylophus vitiensis	Plate 12C: G.R. Zug [captive]
Conolophus marthae	Plate 13E: Thomas H. Fritts [Galápagos, Isabela]

Conolophus pallidus

Plate 13C: Robert P. Reynolds [Galápagos, Sante Fé]

Conolophus subcristatus

Plate 13D: Robert P. Reynolds [Galápagos, Plaza]

Iguana iguana

Plate 12D: Kevin M. Enge [Florida, Hollywood]

Phrynosomatidae

Urosaurus auriculatus

Plate 14F: Bayard Brattstrom [Revillagigedo, Socorro]

Tropiduridae

Microlophus albemarlensis

Plate 14A: Kevin de Queiroz [Galápagos, Santa Cruz] male

Microlophus albemarlensis

Plate 14B: Kevin de Queiroz [Galápagos, Santa Cruz] female

Microlophus bivittatus

Plate 14C: Kevin de Queiroz [Galápagos, San Cristóbal]

Microlophus delanonis

Plate 14D: Robert P. Reynolds [Galápagos, Española]

Microlophus grayii

Plate 14E: Kevin de Queiroz [Galápagos, Floreana]

Eugongylidae

Carlia ailanpalai

Plate 22A: G.R. Zug [Guam]

Cryptoblepharus eximius

Plate 16A: Robert N. Fisher [Fiji, Viti Levu]

Cryptoblepharus eximius

Plate 16B: Christopher C. Austin [Fiji, Taveuni]

Cryptoblepharus novohebridicus

Plate 16C: Alison M. Hamilton [Vanuatu, Tanna]

Cryptoblepharus poecilopleurus

Plate 16D: G.R. Zug [Hawaii, Kaui'i]

Cryptoblepharus rutilus

Plate 16E: Christopher C. Austin [Palau]

Emoia adspersa

Plate 23A: Robert N. Fisher [Samoa]

Emoia aneityumensis

Plate 19A: Alison M. Hamilton [Vanuatu, Anatom]

Emoia arnoensis

Plate 22B: Donald W. Buden [Federated States of Micronesia]

Emoia atrocostata	Plate 23B: Björn Lardner [Guam]
Emoia atrocostata	Plate 23C: Christopher C. Austin [Palau]
Emoia boettgeri	Plate 22C: Donald W. Buden [Federated States of Micronesia, Pohnpei]
Emoia caeruleocauda	Plate 17A: G.R. Zug [Guam] adult-striped
Emoia caeruleocauda	Plate 17B: Elijah Wostl [Guam] adult-unicolor
Emoia caeruleocauda	Plate 17C: G.R. Zug [Fiji, Taveuni] juvenile
Emoia campbelli	Plate 21A: Paddy Ryan [Fiji, Viti Levu]
Emoia concolor	Plate 21B: G.R. Zug [Fiji, Taveuni]
Emoia cyanogaster	Plate 21D: G.R. Zug [Vanuatu, Efate]
Emoia cyanura	Plate 17D: A.H. Whitaker [Samoa, Upolu]
Emoia erronan	Plate 19B: Alison M. Hamilton [Vanuatu, Futuna]
Emoia impar	Plate 17E: A.H. Whitaker [Cook Islands, Rarotonga]
Emoia jakati	Plate 18A: Mike McCoy [Bougainville]
Emoia lawesii	Plate 23D: Robert N. Fisher [Samoa]
Emoia mokolahi	Plate 20A: G.R. Zug [Tonga, 'Eua]
Emoia mokosariniveikau	Plate 21C: G.R. Zug [Fiji, Vanua Levu]
Emoia oriva	Plate 20B: Paddy Ryan [Fiji, Rotuma]
Emoia nigra	Plate 22D: G.R. Zug [Samoa, Savaii]
Emoia nigromarginata	Plate 19C: G.R. Zug [Vanuatu, Efate]
Emoia parkeri	Plate 23E: G.R. Zug [Fiji, Viti Levu]
Emoia ponapea	Plate 23F: Christopher C. Austin [Federated States of Micronesia, Pohnpei]
Emoia rufilabialis	Plate 18B: Mike McCoy [Solomon, Santa Cruz Island] female
Emoia rufilabialis	Plate 18C: Mike McCoy [Solomon, Santa Cruz Island] male
Emoia samoensis	Plate 20C: A.H. Whitaker [Samoa]
Emoia sanfordi	Plate 19D: G.R. Zug [Vanuatu, Espiritu Santo]
Emoia slevini	Plate 20F: Elijah Wostl [Guam, Cocos Island]

Emoia taumakoensis	Plate 18D: Mike McCoy [Solomon, Taumako] female
Emoia taumakoensis	Plate 18E: Mike McCoy [Solomon, Taumako] male
Emoia tongana	Plate 21E: G.R. Zug [Samoa, Savaii]
Emoia trossula	Plate 20E: G.R. Zug [Fiji, Taveuni]
Emoia tuitarere	Plate 20D: G.R. Zug [Cook Islands, Rarotonga]
Eugongylus albofasciolatus	Plate 24B: Christopher C. Austin [Federated States of Micronesia, Pohnpei]

Lygosomidae

Caledoniscincus atropunctatus	Plate 24A: A.H. Whitaker [New Caledonia]
Lamprolepis smaragdina	Plate 19E: Margie Falanruw [Federated States of Micronesia, Yap]
Lampropholis delicata	Plate 26E: A.H. Whitaker [Australia]
Leiolopisma alazon	Plate 24E: Robert N. Fisher [Fiji, Ono-i-Lau]

Mabuyidae

Eutropis multicarinata	Plate 25A: Donald W. Buden [Federated States of Micronesia, Yap]

Sphenomorphidae

Lipinia leptosoma	Plate 24C: Christopher C. Austin [Palau]
Lipinia noctua	Plate 24D: Robert N. Fisher [unknown]
Prasinohaema virens	Plate 21F: Christopher C. Austin [Papua New Guinea]
Sphenomorphus scutatus	Plate 25B: Elijah Wostl [Palau, Koror]
Sphenomorphus solomonis	Plate 25C: Michael McCoy [Solomon Islands, Guadalcanal]

Varanidae

Varanus "indicus"	Plate 26A: Elijah Wostl [Guam, Cocos Island]
Varanus "indicus"	Plate 26B: Margie Falanruw [Federated States of Micronesia, Yap]

Other Lizards, Mostly Aliens

Chamaeleo calyptratus — Plate 26C: Kevin M. Enge [Florida]

Trioceros jacksonii — Plate 26D: John H. Tashjian [captive]

Diploglossus millepunctatus — Plate 25D: Mateo Lopez-Victoria [Malpelo]

Reptilia: Squamata (Snakes)

Typhlopidae

Ramphotyphlops acuticaudus — Plate 27A: Christopher C. Austin [Palau]

Ramphotyphlops adocetus — Plate 27D: Margie Falanruw [Federated States of Micronesia, Ant Atoll]

Ramphotyphlops braminus — Plate 27B: G.R. Zug [Myanmar, Chatthin]

Ramphotyphlops depressus — Plate 27C: Mike McCoy [Bouganville]

Ramphotyphlops hatmaliyeb — Plate 27E: Margie Falanruw [Federated States of Micronesia, Ulithi]

Boidiae

Candoia bibroni — Plate 28A: G.R. Zug [Samoa, Savaii]

Candoia bibroni — Plate 28B: G.R. Zug [Fiji, Kadavu]

Candoia paulsoni — Plate 28C: Mike McCoy [Solomon Islands, Guadalcanal]

Candoia superciliosa — Plate 28D: Christopher C. Austin [Palau]

Colubridae

Alsophis biseralis — Plate 29A: Heidi Snell [Galápagos]

Alsophis biseralis — Plate 29B: Heidi Snell [Galápagos, Santiago]

Antillophis slevini — Plate 29C: Aaron Kessler/T.Fritts [Galápagos, Fernandina]

Antillophis steindachneri — Plate 29D: Heidi Snell [Galápagos, Santiago]

Boiga irregularis — Plate 30A: Elijah Wostl [Guam]

Dendrelaphis salomonis — Plate 30B: Mike McCoy [Solomon Islands, Guadalcanal]

Dendrelaphis striolatus — Plate 30C: Gregory Pregill [Palau, Koror]

Philodryas hoodensis — Plate 29E: Heidi Snell [Galápagos]

Elapidae

Hydrophis coggeri Plate 31A: H. Cogger [Ashmore Reef]

Laticauda colubrina Plate 31B: Christopher C. Austin [Palau]]

Laticauda frontalis Plate 31C: H. Cogger [Vanuatu, Espiritu Santo]

Laticauda laticaudata Plate 31D: H. Cogger [New Caledonia, Noumea]

Ogmodon vitianus Plate 30E: G.R. Zug [Fiji, Viti Levu]

Pelamis platura Plate 31E: H. Cogger [Ashmore Reef]

Homalopsidae

Cerberus dunsoni Plate 30D: Christopher C. Austin [Palau, Koror]

Reptilia: Testudines

Chelidae

Chelodina longicollis Plate 33A: H.Cogger [Australia]

Cheloniidae

Caretta caretta Plate 32A: Paddy Ryan [Bahamas]

Chelonia mydas Plate 32B: G.R. Zug [Hawaii, O'ahu]

Eretmochelys imbricata Plate 32C: A.H. Whitaker [Vanuatu, Efate]

Lepidochelys olivacea Plate 32D: Margie Falanruw [Federated States of Micronesia, Yap]

Dermochelyidae

Dermochelys coriacea Plate 32E: Thomas E.J. Leuteritz [Gabon, Pongara]

Chelydridae

Chelydra serpentina Plate 33B: Robert G. Tuck [Maryland]

Emyidae

Trachemys scripta Plate 33E: Steve W. Gotte [Texas, Pecos River]

Geoemydidae

Mauremys reevesii   Plate 33C: Carl H. Ernst [captive]

Mauremys sinensis   Plate 33D: Indraneil Das [captive]

Testudinidae

Chelonoidis darwini   Plate 34A: Robert P. Reynolds [Galápagos, Santiago]

Chelonoidis ephippium   Plate 34B: Robert P. Reynolds [Galápagos, Pinzón]

Chelonoidis guentheri   Plate 34D: Thomas Fritts [Galápagos, Isabela]

Chelonoidis porteri   Plate 34E: Robert P. Reynolds [Galápagos, Santa Cruz]

Chelonoidis vandenburghi   Plate 34C: Robert P. Reynolds [Galápagos, Isabela]

Chelonoidis vicina   Plate 34F: Robert P. Reynolds [Galápagos, Isabela]

Trionychidae

Apalone spinifera   Plate 35A: Steve W. Gotte [Alabama]

Palea steindachneri   Plate 35B: Carl H. Ernst [captive]

Pelodiscus sinensis   Plate 35C: Hans-Dieter Phillip [captive]

Pelodiscus sinensis   Plate 35D: Robert T. Zappalorti [captive]

Reptilia: Crocodilia

Crocodylidae

Crocodylus porosus   Plate 35E: Daniel Moen [Australia, Northern Territory]

MAPS

Source information for each map is arranged sequentially by map number.

Map 1. Pacific Ocean base map from the research collection of G.R. Zug. Illustrator, T. Britt Griswold. Labeling by G.R. Zug.

Maps 2–5. These derive from the preceding base map, with labeling by G.R. Zug.

FIGURES

Source information for each figure is presented sequentially by figure number. I have drawn freely from the figures published in my research articles. The research article illustrations were created during my tenure as a federal employee of the Smithsonian Institution. Molly D. Griffin and Ted Kahn prepared new illustrations, specifically for this guide, under contract to G.R. Zug.

Figure 1. Illustrator, Ted Kahn.

Figure 2. Illustrator, Ted Kahn.

Figure 3. From Fig. 2 in G.R. Zug, 1991, *Bishop Museum Bulletins in Zoology*, no 2. Illustrator, Esta L. Johnston.

Figure 4. Illustrator, Ted Kahn.

Figure 5. From Figs. 2 and 4 in G.R. Zug, 1991, *Bishop Museum Bulletins in Zoology*, no 2. Illustrator, Esta L. Johnston.

Figure 6. From Figs. 2 and 5 in G.R. Zug, 1991, *Bishop Museum Bulletins in Zoology*, no 2. Illustrator, Esta L. Johnston.

Figure 7. From Fig. 5 in G.R. Zug and B. Moon, 1995, Herpetologica 51(1). Illustrator, Kate Spencer.

Figure 8. All illustrations modified from Fig. 1 in W. Brown, 1976, *Occasional Papers of the California Academy of Science* 126.

Figure 9. Modified from Plate XII, Fig. 4a in G.A. Boulenger, 1885, *Catalogue of the Lizards in the British Museum (Natural History)*, Second Edition. Vol.1, London. Printed by order of the trustees.

Figure 10. *Gonatodes* and *Phyllodactylus* digits modified from Plate V, Fig.1a, and Plate VII, Fig. 2a, in G.A. Boulenger, 1885, *Catalogue of the Lizards in the British Museum (Natural History)*, Second Edition. Vol. 1, London. Printed by order of the trustees. *Sphaerodactylus* digit modified from Fig. 114e of J.C. Lee, 1995, *The Amphibians and Reptiles of the Yucatan Peninsula*. Ithaca: Comstock Publishing Associates.

Figure 11. Modifed from Figs. 4 and 6 in E.H. Taylor, 1942, *University of Kansas Science Bulletin* 28 (6).

Figure 12. *Anolis carolinensis* from Fig. 16 in E.D. Cope, 1899, *The Crocodilians, Lizards, and Snakes of North America: Report of the U.S. National Museum for 1898*. Washington, DC: Smithsonian Institution. *Anolis equestris* after de la Sagra, 1854, in A. Fläschendräger and L.C.M. Wijffels, 1996, "Anolis." In *Biotop und Terrarium*. Münster: Natur u. Tier Verlag.

Figure 13. *Carlia fusca*, from the research collection of G.R. Zug. Illustrator, T. Britt Griswold. *Eutropis multicarinata* from Plate XI, Fig. 2, in G.A. Boulenger, 1885, *Catalogue of the Lizards in the British Museum (Natural History)*, Second Edition. Vol. 3, London. Printed by order of the trustees.

Figure 14. *Cryptoblepharus eximius* from Fig. 7 in G.R. Zug, 1991, *Bishop Museum Bulletins in Zoology*, no 2. *Emoia adspersa* from the research collection of G.R. Zug. Illustrator, Esta L. Johnston.

Figure 15. From Fig. 8 in G.R. Zug, 1991, *Bishop Museum Bulletins in Zoology*, no 2. Illustrator, Esta L. Johnston.

Figure 16. Illustrator, Ted Kahn.

Figure 17. From Fig. 14 and 15 in G.R. Zug, 1991, *Bishop Museum Bulletins in Zoology*, no 2. Illustrator, Esta L. Johnston.

Figure 18. Illustrator, Ted Kahn.

Figure 19. Modified from Fig. 14 in M.A. Smith, 1943, *Fauna of British India: Reptilia and Amphibia*. Vol. III – Serpentes. London: Taylor & Francis.

Figure 20. *Candoia superciliosa* [Palau]. Illustrators, Molly Dwyer Griffin and Ted Kahn. *Candoia bibroni*, Illustrator, Ted Kahn.

Figure 21. Illustrator, Ted Kahn.

Figure 22. Illustrator, Molly Dwyer Griffin.

Figure 23. From Figures 328, 329, 338, 339, 355, and 356 in L. Stejneger, 1907, "Herpetology of Japan and Adjacent Territory," *United States National Museum Bulletin 58*.

Figure 24. Correction of head illustration of *Ogmodon vitiensis* after Fig. 4, 4a in W. Peters 1864, *Monatsberichte der königlich Akademie der Wissenschaften zu Berlin,* 1864 (April). Illustrator, Ted Kahn.

Figure 25. Illustrator, Molly Dwyer Griffin.

Figure 26. Illustrator, Ted Kahn.

Figure 27. *C. serpentina*, from E. Bobbe in Fig. 2b, H. Wermuth and R. Mertens, 1961, *Schildkröten Krokodile Brückenechsen*. Jena: Gustav Fischer Verlag. *M. reevesii*, from E. Bobbe in Fig. 34c, H. Wermuth and R. Mertens, 1961, *Schildkröten Krokodile Brückenechsen*. Jena: Gustav Fischer Verlag.

Figure 28. *M. reevesii*, modified from C. Pope, 1935 in Fig. 34b, H. Wermuth and R. Mertens, 1961, *Schildkröten Krokodile Brückenechsen*. Jena: Gustav Fischer Verlag. *M. sinensis*, modified from C. Pope 1935 in Figure 103b, H. Wermuth and R. Mertens, 1961, *Schildkröten Krokodile Brückenechsen*. Jena: Gustav Fischer Verlag. *T. s. elegans*, from photograph, G.R. Zug.

Figure 29. Illustrator, Molly Dwyer Griffin.

Figure 30. Illustrator, Molly Dwyer Griffin.

Index of Common English Names

Index of Scientific Names

Alsophis biseralis, 47, 222, 283, 294, Plate 29

Amblyrhynchus cristatus, 46, 134, 135, 280, 290, Plate 13

Anolis agassizii, 45, 127, 128, 280, 290, Plate 15

Anolis carolinensis, 11, 12, 15, 16, 19, 128, 129, 153, 157, 280, 290, Plate 15

Anolis equestris, 12, 128, 130, 131, 280, 290, Plate 15

Anolis sagrei, 12, 131, 132, 280, 290, Plate 15

Anolis townsendi, 45, 132, 133, 280

Antillophis slevini, 47, 222, 223, 283, 294, Plate 29

Antillophis steindachneri, 47, 223, 284, 294, Plate 29

Apalone spinifera, 13, 264, 265, 285, 296, Plate 35

Boiga irregularis, 16, 30, 87, 223, 224, 284, 294, Plate 30

Brachylophus bulabula, 33, 35, 135, 136, 280, 290, Plate 12

Brachylophus fasciatus, 35, 36, 136, 280, 290, Plate 12

Brachylophus vitiensis, 35, 137, 138, 280, 290, Plate 12

Caledoniscincus atropunctatus, 33, 191, 192, 282, 293, Plate 24

Candoia bibroni, 30, 33, 34, 35, 37, 38, 218, 219, 221, 283, 294, Plate 28

Candoia paulsoni, 30, 219, 220, 283, 294, Plate 28

Candoia superciliosa, 19, 220, 221, 283, 294, Plate 28

Caretta caretta, 11, 13, 29, 33, 36, 37, 39, 41, 241, 243, 284, 295, Plate 32

Carlia ailanpalai, 15, 16, 20, 21, 152, 153, 154, 184, 281, 291, Plate 22

Carlia fusca, 153

Carlia tutela, 19, 154, 281

Cerberus dunsoni, 19, 238, 284, 295, Plate 30

Cerberus rynchops, 237, 238

Chamaeleo calyptratus, 12, 209, 283, 294, Plate 26

Chelodina longicollis, 41, 239, 240, 284, 295, Plate 33

Chelonia mydas, 11, 13, 15, 16, 17, 20, 21, 22, 23, 24, 25, 26, 27, 28, 29, 30, 31, 33, 36. 37, 38, 39, 41, 42, 43, 44, 45, 47, 48, 242, 243, 284, 295, Plate 32

Chelonoidis abingdonii, 47, 256, 285

Chelonoidis becki, 47, 257, 258, 285

Chelonoidis chathamensis, 47, 257, 258, 285

Chelonoidis darwini, 47, 256, 258, 285, 296, Plate 34

Chelonoidis duncanensis, 259

Chelonoidis ephippium, 47, 258, 259, 285, 296, Plate 34

Chelonoidis guentheri, 47, 259, 260, 285, 296, Plate 34